Seed Science

Seed Science

Cory Young

Larsen & Keller
www.larsen-keller.com

Seed Science
Cory Young
ISBN: 978-1-64172-603-0 (Hardback)

© 2022 Larsen & Keller

Larsen & Keller

Published by Larsen and Keller Education,
5 Penn Plaza,
19th Floor,
New York, NY 10001, USA

Cataloging-in-Publication Data

Seed science / Cory Young.
   p. cm.
Includes bibliographical references and index.
ISBN 978-1-64172-603-0
1. Seeds. 2. Seed technology. 3. Seeds--Development.
4. Agriculture. I. Young, Cory.
SB117 .S44 2022
631.521--dc23

For more information regarding Larsen and Keller Education and its products, please visit the publisher's website www.larsen-keller.com

TABLE OF CONTENTS

This book aims to help a broader range of students by exploring a wide variety of significant topics related to this discipline. It will help students in achieving a higher level of understanding of the subject and excel in their respective fields. This book would not have been possible without the unwavered support of my senior professors who took out the time to provide me feedback and help me with the process. I would also like to thank my family for their patience and support.

The fertilized and mature ovules of plants which are used for sowing are known as seeds. The plants which use seeds for their reproduction are known as spermatophytes. The branch of agricultural sciences that deals with the study of the structure of seeds is known as seed science. It also studies the development of seeds from a fertilized egg cell to the emergence of a new plant. Some of the focus areas of this branch of science are- the different types of germination, seed storage and seed ecology. It is also involved in studying the different methods and techniques which are involved in the sowing of seeds. This book provides comprehensive insights into the field of seed science. The ever growing need of advanced technology is the reason that has fueled the research in this field in recent times. Coherent flow of topics, student-friendly language and extensive use of examples make this book an invaluable source of knowledge.

A brief overview of the book contents is provided below:

Chapter – Introduction to Seed Science

An embryonic plant which is encapsulated within a protective external covering is known as a seed. They are generally classified into four classes, namely, breeder seed, foundation seed, registered seed and certified seed. The topics elaborated in this chapter will help in gaining a better perspective about the different types of seeds as well as their morphological aspects.

Chapter – Seed Ecology

The study of ecological strategies which are used by plants in order to ensure their reproduction through seeds is known as seed ecology. The plants which produce seeds are known as spermatophytes. These are further sub-divided into two broad categories, namely, gymnosperms and angiosperms. This chapter has been carefully written to provide an easy understanding of these varied elements of study within seed ecology.

Chapter – Seed Germination

The process through which a plant grows from a seed is known as germination. The germination of plant can take place underground or above the ground. Underground germination is known as hypogeal germination and above ground germination is known as epigeal germination. All these diverse aspects of germination have been carefully analyzed in this chapter.

Chapter – Sowing Techniques and Methods

The process of planting seeds by scattering them on or inside the earth is known as sowing. Some of the techniques which are used to sow seeds are dibbling, aerial seeding, precision seeding and drilling. All the diverse methods and techniques related to sowing have been carefully analyzed in this chapter.

Chapter – Diverse Aspects of Seed Science

The branch of science which is involved in the processing, testing and conservation of food and seeds is known as seed science. Some of the areas which are under focus in this discipline are seed testing, seed pelleting and seed storage. The chapter closely examines these key concepts and techniques of seed science to provide an extensive understanding of the subject.

Cory Young

Introduction to Seed Science

An embryonic plant which is encapsulated within a protective external covering is known as a seed. They are generally classified into four classes, namely, breeder seed, foundation seed, registered seed and certified seed. The topics elaborated in this chapter will help in gaining a better perspective about the different types of seeds as well as their morphological aspects.

SEEDS

Seeds are many things, but everything about seeds—their numbers and forms and structures—has a bearing on their main purpose, to insure continuing life. Seeds are containers of embryonic plants, the embryos of a new generation. Seeds are borne by two great and different classes of plants.

One group, less highly developed than the other, produce "naked" seeds that develop from "naked" ovules. In plants of the more highly developed and much larger class, the ovule and the seed develop within an ovary, the seed vessel. The ovary is the part of the flower that contains the ovule with its egg, or female sex cell. The ovary later becomes a fruit with the developed ovule or ovules—seeds— inside. This group of plants we call angiosperms, a word that means vessels for seeds.

Plants of the other group, the gymnosperms, the "naked seed" plants, have no ovaries, no flowers, and no fruits, although they do have seeds. Gymnosperms include the cone-bearing trees, the conifers. Their seeds are borne in pairs at the bases of the scales of the cones.

Deep within the ovary of the mother flower (or between the scales of a seed cone) lies the ovule, which contains an embryo sac and its tiny egg. The egg must be fertilized by a sperm cell from a pollen tube before it can start to develop into an embryo and so perpetuate the parent's life.

Along with the embryo there develops a special store of food, the embryo's own special "formula" or diet for its use after it is separated from its mother plant.

Every seed contains carbohydrates, proteins, fats, and minerals to nourish the embryonic plant within. The nature and proportions of each of them differ among the many kinds of seeds. Some seeds, like corn, are predominantly starchy. Seeds of flax and sunflower are oily or fat. Others, such as peas and beans, are notable for their high content of protein.

Some seeds (such as the seeds of orchids, which are like specks of dust) contain only tiny bits of stored food because they are so small. Large seeds may contain a billion times more food than the smallest ones. Some kinds of seed have most of their reserve supplies packed inside their seed leaves. Others have it packed in tissues developed from the embryo sac, called endosperm, or from the cells of the ovule that surrounded the embryo sac.

The seed usually is well protected through its development. This protection differs greatly among different kinds in degree and in the way it is provided. The ovary and the tissues that are attached to it become the fruit of the plant. The seeds (formed by ovules in the ovary) of plants having large or fleshy fruits are deeply protected therefore so that we never see them before maturity unless we open the fruit to find them.

Although the seeds of gymnosperms are said to be naked, they nearly always have some protection during development. The seeds of the pine tree and other conifers, for example, are hidden at the bases of the scales of the cone. The cone scales of some pines separate to release the seeds as soon as they are mature. Others remain closed for years.

The fruit tissues that enclose some seeds are scanty and are attached to the coat of the seed. A kernel of corn, for example, is more than a seed—it is a one-seeded fruit. The kernel is nearly all seed, but a thin layer of ovary tissue surrounds the seed and has grown together with the seedcoat in such a way that we can hardly see the tissue.

Many structures that we call seeds are actually fruits. Most of them, such as the fruits of the cereals and other grasses, lettuce, and spinach, contain only one seed. Members of the carrot family produce two-parted fruits, each with one seed. Some fruits, such as those of beets, have one or several seeds. Botanists identify the various types of fruits and give them specific names, but our purpose here is served if we deal with the small, dry, one- or fewseeded fruits, which we are accustomed to plant like seeds, as though they were seeds.

Seeds of some species develop in the mother plant with amazing speed. Some others are surprisingly slow. A chickweed plant that is pulled from the garden and thrown aside at the time its flowers first open may form some seeds before it withers and dies.

Most familiar plants form their seeds during a period of several days to a few weeks following pollination. Pine trees, however, take 2 to 3 years to mature their seeds. The fruit of the sea palm is said to need 7 to 10 years to mature.

Another aspect of the survival of plants is that the seed-bearing species can be perpetuated in two ways. One, which we have been discussing, is sexual—that is, by means of seeds, which develop from fertilized egg cells. The other is asexual, or vegetative, as we usually say, by means of such parts as buds, pieces of root, and pieces of stem with attached buds, bulbs, and tubers.

The seeds of some plants—like potatoes, cultivated tree fruits, grapes, berries, and many ornamental garden plants—do not come true to variety. Their seeds therefore are worthless for perpetuating the varieties we plant in gardens and orchards.

For them, we must use vegetative propagation. We can grow apple trees, grapevines, potatoes, or strawberry plants from seeds, but the plants and their fruits (or tubers) will be unlike those of the varieties that produced the seeds.

That is because most seeds, as we have seen, develop after the union of male and female reproductive cells. The seeds perpetuate the hereditary characteristics contributed by both the male and female cells. Seeds of plants like potatoes, apples, pears, and tulips fail to come true to variety because their sex cells carry random assortments of mixed-up sets of characters. Among the offspring of the numberless chance unions that occur in such plants, hardly any two are alike. The plants from seeds of most species come reasonably true to variety if precautions are taken to keep the pollen of undesired types from reaching the flowers of desired types.

We must note a rare exception. A few kinds of plants, such as some species of grasses and of Citrus^ produce asexual seeds, whose embryos develop entirely from cells of the ovule outside the egg apparatus. No fertilization of an egg cell is involved. There is no mixture of characters from pollen cells with those of the mother cells. The embryo is formed entirely from mother plant cells and therefore is identical with the mother plant in its hereditary makeup. Such asexual seeds, therefore, come true to variety and afford the unusual opportunity of accomplishing "vegetative" propagation by means of seeds. Except for such rare instances, however, seed propagation means sexual propagation, and asexual or vegetative propagation means propagation by some means other than seeds.

Plants that do come true to variety from seed of sexual origin can also be propagated asexually from stem cuttings or other appropriate parts of the plant under favorable conditions.

A prohibitive amount of work would be required for the vegetative propagation of the billions upon billions of such plants that v/e need to grow every year. An even greater obstacle is that there is no feasible way to keep these 'Vegetative" plants alive through periods of great cold, drought, or flood, if such plants are killed before they produce seed, that is the end of their line.

The kinds and varieties of plants that fail to produce viable seeds—that is, seeds that can grow or develop—must be perpetuated by asexual means. There is no other way.

Such diverse plants as certain grasses, bananas, and garlic produce no seeds, but each has an asexual feature (a vegetative structure) by which it can be multiplied.

Sometimes, for a particular reason, growers resort to vegetative propagation of a kind of plant that is normally grown only from seeds.

Small farmers in the hills of Vietnam grow cabbage year after year without the use of seeds. The climate there is not cool enough at any time to induce flowering and seed production in cabbage, and the farmers cannot import seeds for each planting. The farmers therefore make cuttings from the stumps of the cabbage plants after the heads are harvested. They plant the pieces of stump, each of which has one or more side buds. Roots soon develop. The buds grow and produce new cabbage plants that will develop heads. The process is repeated for each crop.

This method of growing cabbage would be impossible where the seasons become too cold, too hot, too wet, or too dry for the continued survival of the vegetative stage of the plant. The enormous numbers of seeds that single plants of some species produce make it feasible to increase seed supplies at almost fantastically rapid rates. Single plants of other species produce few seeds, and the rates of increase are ploddingly slow.

One tobacco plant may produce as many as 1 million seeds. The average is about 200 thousand seeds. The garden pea plant produces a few dozen seeds at best. The possible rate of spread of some plants over an area by seeds therefore is astronomical. With other plants, the rate is modest or slow. Even the relatively slow rates of seed increase among annual and biennial plants are fast and easy, compared to most vegetative propagation.

Species and varieties of hardy perennial plants that spread by runners (creeping stems above the soil surface), stolons (creeping stems below the soil surface), bulbs (arrangements of fleshy leaf bases on a drastically shortened stem), and tubers (greatly thickened underground stems), are especially adapted to survival for long periods without depending on seeds, although they may also produce seeds. Seeds of these kinds of plants often do not come true to variety.

Plants are able to spread naturally only very slowly if seeds are absent. They can only creep. Their vegetative structures do not fly on the wind, float on the water, or ride on animals to distant sites as easily as seeds do. Vegetative reproductive parts may be torn from parent plants by animals or by storms and later may take root after being carried some distance. Vegetative spread nevertheless is slow and cumbersome in nature, compared to spreading by seeds.

SEEDS are the protectors as well as the propagators of their kinds. Thousands of kinds of plants have evolved in such ways that they cannot survive, even in the regions where they are best adapted, if they produce no seeds.

Seeds of most plants are the very means of survival of the species. They carry the parent germ plasm, variously protected against heat, cold, drought, and water from one growing season that is suitable for growth of the species to the next.

Most kinds of seeds will live considerably longer than the time from one growing season to the next if their surroundings are not too extreme for their respective characteristics. Some seeds normally keep alive under natural conditions above ground only a year or two. Others can keep alive for a score of years or more. A few, such as the seeds of silver maple, remain viable only a few days if they are not kept moist and cool.

Some kinds can survive deep burial in the soil, dry or moist, for lo to 20 years or longer. In one famous experiment, started in 1902, J. W. T. Duvel, of the Department of Agriculture, placed some seeds in soil in flowerpots, so he could find them later. He then buried the pots and all. At intervals he dug up the pots, recovered the seeds, and then planted them under favorable conditions for germination. More than 50 of 107 species tested were viable after 20 years. Many weed seeds remain viable for a very long time if they are buried deeply.

Seeds of common evening-primrose and mullein have been known to remain viable after 70 years in soil.

Most crop seeds keep best for one or a few years when they are stored in a dry place. Exposure to warm, moist air shortens their life. Repeated wetting or submergence in water soon kills most of them. Seeds of plants that grow in water, on the other hand, are not soon harmed by water.

Onion seeds kept in a warm, humid place will lose their life in a few months. When they are well dried and sealed in glass, they remain viable more than a dozen years at room temperature. If seeds are relatively dry, most kinds will tolerate for years extreme cold that would quickly kill their parent plants.

Most seeds also tolerate prolonged hot weather if they are dry. Seeds of muskmelon have produced good plants in the field after storage in a hot, dry office for 30 years. Seeds of Indian-mallow, a common weed, have germinated after 70 years of dry storage. Seeds of Mimosa, Cassia and some other genera have germinated after being kept in a herbarium more than 200 years.

The seeds of Lagenaria a gourd, are not harmed by the immersion of the fruits in sea water for a year, long enough for the fruits to float across an ocean. Water may enter the fruits and wet the seeds. Lotus seeds estimated to be 800 to 1,200 years old have germinated.

The stories, however, about the finding of viable seeds 2 thousand to 3 thousand years old in Egyptian tombs are not true. Viable barley seeds found in the wrappings of a mummy were traced to the new straw in which the mummy was packed for shipment to a museum. Viable seeds of corn, squash, and beans found in caves and ancient ruins of cliff dwellings had not lain there for hundreds of years—pack rats or other creatures had carried them in not long before the archeologists found them.

The long "storage" life of the embryo within the seed not only helps insure survival of the species, it makes possible the distribution or spread of the species over long distances, either in the wild or by the agency of man.

Viable seeds probably are never completely inactive. Vital processes go on as a seed awaits conditions favorable for germination and plant growth. If we knew how to arrest or suspend all these processes completely, it would be possible theoretically to retain viability indefinitely. We do not know how to do that.

Activity within the seed may be so low that we cannot measure it by any known method. In time, however, if the seed does not encounter conditions that will permit it to grow, unidentified substances become exhausted or they deteriorate, and germinating power is lost. The seed dies. Warmth and moisture hasten the exhausting life processes and shorten the life of the seed. Dryness and cold slow down activities, conserve vital substances, and protect the delicately balanced systems within the seed.

Seeds possess remarkably complex and effective protective mechanisms that help insure survival. Consider a tender plant that grows in a region of sharply different seasons and matures its seeds and drops them to the ground while the weather is still favorable for growth, If those seeds grow promptly, the new plant surely will be killed when winter comes. In such situations, seeds that grow promptly are wasted because they fail to perpetuate the parents.

Many seeds therefore have a rhythm of ability to grow that coincides with the rhythm of the seasons. They have a delayed-action mechanism, a natural timeclock, which insures that the seeds will remain dormant until another growing season rolls around—a season long enough to permit another generation of seeds to mature.

Many kinds of seeds remain dormant—fail to grow upon planting—for a while after separation from the mother plant. The length of the dormancy and the nature of the delaying mechanism differ greatly among species and varieties.

Dormancy that is due to waterresistant ("hard") seedcoats may last for years, until enough water has soaked into the seed for it to germinate. Tiny nicks or scratches in the seedcoat will permit water to enter, thus breaking the dormancy. Natural abrasion of the seeds—by the freezing and thawing of soil or by their movement among rock particles by water—permits water to enter the seed. Hard seeds of crop plants are abraded artificially to induce germination. Dormancies due to some other mechanisms may be overcome less easily.

Some dormant seeds, before they will grow, must go through a long period of cool temperature while they are moist. They must go through conditions that simulate a cold, moist soil during autumn or winter. The rhythm of the seasons must be simulated in the environment of the seeds if they are to grow.

Some seeds lie dormant, although they are in moist soil, until they are exposed to light. Certain weed seeds never germinate deep below the soil surface, but grow quickly after they are brought to the surface when the soil is worked.

Still other seeds fail to grow soon after separation from the mother plant because they are immature. Structural developments or chemical processes, or both, must be completed before they can grow. The naked seed of the ginkgo tree drops to the ground in the autumn long before its embryo is fully grown. The embryo must continue its development for many months, nourished by the foods stored around it, before it is mature enough to break out of the seedcoat and grow.

Some seeds in a nondormant state after harvest can be pushed into a dormant state. Upon exposure to unfavorably warm and moist conditions, some varieties of lettuce seed become dormant, although they are capable of germinating under favorable conditions. It is as though their growth processes recoiled, or went into reverse, in the face of a situation that would be unfavorable for the plants developed from those seeds.

Witchweed, a semiparasitic seedbearing plant, has an unusual survival device. Witchweed is a parasite on many species of crop plants and weeds. Its almost microscopic seeds may lie dormant in the soil for many years if no suitable stimulator plant grows close to them. When the root of a stimulator plant grows close to them, some substance from the root causes the seeds to germinate. The young witchweed plants promptly become parasitic on the roots of any host that caused the seeds to germinate. If a nonhost should cause the seeds to germinate in the absence of a host, the witchweed seedlings die.

Many species of plants are widespread because their seeds are great travelers. Besides the special features that insure perpetuation of their respective species, plants have other features for spreading the species as far and wide as they are able to grow.

Most of the familiar structures that aid in the natural transport of seeds involve fruits rather than the seeds alone.

The windblown dandelion and thistle "seeds" are one-seeded fruits, called achenes. To each is attached a feathery pappus that serves as a sail and a parachute.

The "sticktights" of Spanish-needle are barbed achenes that catch in the coats of animals and people to be carried afar.

A "tickseed" of the beggarweed plant is a one-seeded fragment of its leguminous pod (fruit). It is covered with minute hooks that make it "sticky."

The flying "seed" of the maple tree is a samara, a one-seeded, one-winged fruit.

The water-resistant seeds in buoyant fruits, large or small, one-seeded or many-seeded, may be carried great distances by water. Coconuts, gourds, and the tiny berries of asparagus are examples.

A few kinds of plants distribute their seeds widely as the entire aboveground part of the mature plant tumbles about over the land, blown by the wind. The Russian-thistle is noteworthy among these tumblcweeds. They sometimes roll for many miles, even over fences and other obstructions, scattering seeds as they go.

Some seeds travel on their own. They need not depend on features of their enclosing fruits or of their mother plants as aids to transportation. The coats of some seeds resemble certain surface features of fruits.

The coat of the pine seed is expanded into a wing, which carries it a short distance. The seed of the milkweed has a tuft of long, silky hairs attached to its coat. The wind carries this seed far. When a seed of flax becomes wet, as by rain, its surface becomes gelatinous. It adheres to whatever touches it and is carried away.

The coats of many seeds are resistant to moisture and to the digestive fluids of animals. If such seeds happen to escape grinding by stones in the crops of birds or by the teeth of animals that eat them, the seeds will pass unharmed through the alimentary tract. Some of them reach congenial soil many miles from where the animal got them.

The seeds of the mesquite tree have been distributed by cattle over millions of acres of formerly good grazing lands in the Southwest. Seedlings of cherry, dogwood, and holly commonly appear where seeds have been dropped by birds far from any parent tree.

Unwanted plants make seeds, too. It seems that undesirable or unwanted plants generally are more prolific seed producers than most of the crop plants that we strive to grow. One investigator estimated that one large tumbling pigweed produces more than 10 million seeds. Many kinds produce loo thousand to 200 thousand seeds per plant.

Weeds are the pests they are partly because they produce so many seeds. More than that, though: The seed and the plants that grow from them have a remarkable capacity for survival. Reproductiveness and survival value have evolved to a high level by natural selection. Seeds of many weeds are such potent survivors and successful travelers that their species have become nuisances over much of the world.

Farmers and gardeners must contend with weeds that arise from seeds. They appear to come suddenly from nowhere—or everywhere. They arrive unnoticed by air, by water, by animals, and by man's devices.

Earlier arrivals have accumulated in the soil and lie there waiting for the husbandman to stir them up to the surface, where they seemingly explode into growth. One investigator recovered 10 thousand to 30 thousand viable weed seeds in patches of soil about a yard square and 10 inches deep. Various kinds of seeds kept dormant a long time by their respective mechanisms persistently produce successive waves of noisome seedlings each time the soil is cultivated.

Weeds thus continue to appear although the grower has not allowed a parent plant to produce seed on the site for years. Survival value many weed seeds will survive in the soil 20 years and some for longer than 70 years.

Many weed seeds have nearly the same size, shape, and density as the crop seeds with which they may become mixed. The complete removal of such weed seeds from crop seeds is difficult and expensive. Weed seeds that contaminate seeds for food or industrial use lower the grade and value of the latter. Weed seeds will continue to pose problems for gardeners, farmers, and processors.

Seeds are an aid in efforts to improve plants. It is said sometimes—incorrectly—that plant improvement can occur only through seeds. Many improved varieties of plants have originated as mutations in vegetative (asexual) cells and have been perpetuated vegetative (asexually). No seeds are involved in such instances, although the plants may be seed producers.

Most purposeful plant "improvements," however, have come about through sexual reproduction and the consequent formation of seeds. Useful variations in hereditary characteristics occur much oftener incidental to sexual propagation than in asexually propagated plants. Plant improvement would be extremely slow and uncertain, indeed, if we could only sit and wait for useful mutations to occur.

Man learned long ago that like begets like among annual and biennial plants. He learned that he could gradually upgrade the plants he grew year after year by saving and planting the seeds produced by the most desirable plants.

We speak of "seed selection" when we really should say "parent selection." Man nevertheless has made productive use of the capacity of seeds to contain, preserve, and perpetuate the properties of selected parent plants. For thousands of years he has been gradually improving plants by the parental characters he has helped to perpetuate.

As research has revealed more and more about how plant characters are inherited, seeds have become an increasingly valuable element in the purposeful modification of plants. Seeds are not only a means of perpetuating and multiplying plants but an essential feature of the most rapid and practicable way of progressively improving them.

As additional desirable plant characters (or more desirable degrees of existing traits) are found in a potential parent plant, they can be combined with other desirable characters in a second potential parent by mating the two. Through the seed that results from this planned union, the desired combination of characters is captured and retained. A step upward is taken. Large progenies generally can be developed rather quickly and inexpensively through the agency of seeds. Large numbers greatly increase the probability of finding truly superior plants for further increase and selection or for further mating with other desirable parent plants.

The compactness and longevity of most seeds enable the plant breeder to store collections of germ plasm in small space at small cost and safely and also to distribute germ plasm readily to distant points. Seeds thus help man in his efforts to produce better plants so that he may live better.

Somewhat different but related aspects of survival have to do with the utility and beauty of seeds—the reasons why we grow them, breed them, and husband their spark of life.

Seeds are the world's principal human food. The American Indians, for example, gathered the seeds of about 250 species in more than 30 families of plants for food. Among these were seeds of more than 50 kinds of pine, nut trees, and oak; more than 40 kinds of grasses, of which corn is most important; 30-odd members of the thistle family (like sagebrush and sunflower), and 20 of the goosefoot family (like saltbush and lambsquarters).

Seeds of the wild species used by the Indians are no less wholesome and nutritious today than in the distant past. Most of those species, however, are less productive or are more trouble to grow or harvest than our present crop plants. Or, the seeds are more trouble to prepare or less attractive to our tastes than the ones we now depend on.

All dry edible seeds are highly concentrated foods. For human food, the seeds of certain grasses, the cereals, are by far the most important group. The seeds of wheat provide more human food than any other plant or animal product, and the seeds of rice are second in importance the world over.

The seeds of rye, barley, corn, sorghums, millets, and oats are also important for human food in different regions of the world. Rye and corn are most important in the Americas and Europe. Rice, wheat, and sorghums are dominant in the Far East. About one-fourth of the supply of human energy in the United States comes from seeds of the cereals; in Europe, about one-half; and in the Far East, about three-fourths. These seeds are relatively easy to grow, harvest, and store. One or another of them can be grown wherever there is any agriculture at all.

Seeds are the raw materials for making a great diversity of important products for use in industry and the arts and for making pharmaceuticals, cosmetics, and alcoholic beverages. Among these various purposes, the oilseeds have the widest range of uses. Millions of tons of both oily and starchy seeds are used every year in this country for products other than food and feed.

Most seeds are objects of beauty of form, proportion, surface, and color. Many seeds are so small that their beautiful features escape us. Many others, although large enough to see easily, are such common, everyday objects that we do not really see them. They are, however, worth our careful observation.

The first and most obvious beauty in most true seeds is in the perfection of their simple forms. Their outlines or silhouettes exhibit endless variations in the curve of beauty. In their entirety, too, we find wide ranges of proportion and different graceful and simple masses that are pleasing to look upon.

The sphere is a thing of beauty in itself, although quite unadorned. Artists have tried to produce nonspherical "abstract" forms that possess such grace and proportion as to call forth a satisfying emotional or intellectual response in the beholder. Some of the nicest ofsuch forms lie all about us, unnoticed, in seeds. The commonest are such basic forms as the sphere, the teardrop, and the ovoid and other variations of the spheroid.

Some of these curving shapes are flattened, elongated, or tapered in pleasing ways. Sometimes they are truncated or sculptured into somewhat rough and irregular form. They may bear prominent appendages, such as wings, hooks, bristles, or silky hairs. Most seeds show a smooth flow of line and surface that is perfection itself.

The details of the surface relief of many seeds are even more beautiful in design and precision than the mass of the seed as a whole. Often you can find minute surface characters of surprising kinds. Surfaces that appear plain and smooth to the unaided eye may be revealed under a good hand lens to have beautiful textures.

Surfaces may be grained or pebbled. They may have ridges like those of Doric columns. They may bear geometric patterns in tiny relief, forming hexagons, as in a comb of honey, or minute dimples may cover the surface. Some irregular surface patterns of surprising beauty sometimes appear under the lens. Surfaces may be a dull matte, or highly glossy, or anywhere in between.

Last but not least in the beauty of seeds are their surface colors. They may be snow white or jet black. The color may be a single solid one, or two or more may be scattered about at random. Colors may form definite patterns that are distinctive and characteristic of the species and variety. The colors may be almost any hue of the rainbow—reds, pinks, yellows, greens, purples—and shades of ivory, tan, brown, steely blue, and purplish black.

Look for all you can see with the unaided eye. Then look at smaller seeds and the surfaces of large seeds with a good hand lens. You will be delighted with what you find.

There is still another beauty, a potential beauty in seeds, that can be seen only as the seed fulfills its ultimate purpose—the production of a new plant possessing its own beauty. This is perhaps the greatest of all: Beauty of general form; grace of stem; the shape, sheen, and color of the leaf; and finally the loveliness of the flower or the lusciousness of a fruit. The cycle is complete, and so we are back to the beauty of a seed.

Seeds are a symbol. They color our language and habits of thought. From prehistoric times man has understood the role of seeds. Ancient languages, ancient cultures, and our own contain many words and concepts based on this understanding. The bible contains several such examples, including the parable of the sower, the use of the word "seed" to mean off-spring or progeny, and references to good and bad seed.

Our language contains both common and technical terms involving "seed," although the meanings are quite unrelated to the subject of plants. The meanings recognize, however, some metaphoric connection in one way or another. "Seed" is a noun, an adjective, and a verb.

Watermen speak of seed oysters, seed pearls, and seed fish. The optician speaks of seeds in glass. The chemist seeds a solution with a crystal to induce crystallization. We speak of the seed of an idea or a plan.

We know a great deal about how seeds are formed and what they do, but we know only a little about why that is so. Many purely practical questions still cannot be answered. We wonder about many features of seeds and their behavior.

Scientists study seeds for two kinds of reasons. It is desirable to learn everything possible about seeds in order that man can produce and use them more efficiently and effectively. Seeds or parts of seeds are especially convenient forms of living material for the study of the fundamentals of life processes in plants.

CLASSES OF SEEDS

There are four generally recognized classes of seeds. They are:

- Breeder seed

- Foundation seed

- Registered seed

- Certified seed

The basis of seed multiplication of all notified varieties/hybrids is the Nucleus seed.

- Nuclear seed: This is the hundred percent genetically pure seed with physical purity and produced by the original breeder/Institute /State Agriculture University (SAU) from basic nucleus seed stock. A pedigree certificate is issued by the producing breeder.

- Breeder seed: The progeny of nucleus seed multiplied in large area as per indent of Department of Agriculture and Cooperation (DOAC), Ministry of Agriculture, Government of India, under supervision of plant breeder/institute/ SAUs and monitored by a committee consisting of the representatives of state seed certification agency, national/state seed corporations, ICAR nominee and concerned breeder. This is also hundred percent physical and genetic pure seed for production of foundation seed. A golden yellow colour certificate is issued for this category of seed by the producing breeder.

- Foundation seed: The progeny of breeder seed produced by recognized seed producing agencies in public and private sector, under supervision of seed certification agencies in such a way that its quality is maintained according to prescribed field ad seed standards. A white colour certificate is issued for foundation seed by seed certification agencies.

- Registered seed: Registered seed shall be the progeny of foundation seed that is so handled as to maintain its genetic identity and purity according to standard specified for the particular crop being certified. A purple colour certificate is issued for this category of seed.

- Certified seed: The progeny of foundation seed produced by registered seed growers under supervision of seed certification agencies to maintain the seed quality as per minimum seed certification standards. A blue colour certificate is issued by seed certification agency for this category of seed.

The foundation and certified seeds can be multiplied at stage I and II, but the reproduction can not exceed three generations after breeder seed.

Difference between Certified Seed and Truthful Labeled Seed.

Certified seed	Truthful labelled seed
Certification is voluntary. Quality guaranteed by certification agency.	Truthful labelling is compulsory for notified kind of varieties. Quality guaranteed by producing agency
Applicable to notified kinds only	Applicable to both notified and released varieties
It should satisfy both minimum field and seed standards	Tested for physical purity and germination
Seed certification officer, seed inspectors can take samples for inspection	Seed inspectors alone can take samples for checking the seed quality

Dicotyledonous and Monocotyledonous Seeds

Dicotyledonous Seeds

1. Gram Seed

The gram seed is more or less rounded at one end and pointed at the other. It is covered by a brown seed coat called testa; the inner whitish coat is the tegmen. At the pointed end of the seed the testa bears a scar called hilum.

It indicates the position of the attachment of the seed to the fruit wall. Near about the hilum there is a very minute opening called micropyle (micro = small; pyle = gate). If a soaked seed is pressed a droplet of water oozes out through this micropyle. The kernel is exposed when the seed coat is removed, which in the gram seed is nothing but the embryo.

If the kernel is pressed the two fleshy cotyledons are found, which remain laterally hinged to the axis of the embryo at a joint called the first node or nodal zone. The lower part of are found, which remain laterally hinged to the axis of the embryo at a joint called the first node or nodal zone.

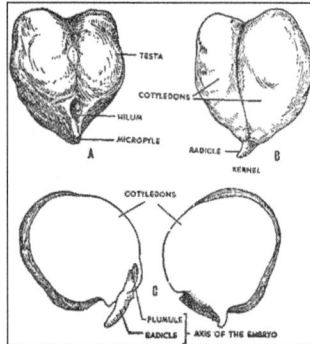

Structure of Gram Seed. A-seed; B-kernel; C-same split open showing parts.

The lower part of the axis lying towards the pointed end is the radicle and the upper part lying between the two cotyledons is the plumule. The plumule has very minute leaves and looks like a feather. Food matters remain stored up in the fleshy cotyledons. So the gram seed is dicotyledonous exalbuminous.

2. Germination of Gram Seeds

When the essential conditions are satisfied, the seed begins to germinate. At first it absorbs water and swells up. The testa ruptures near about the micropyle and the radicle is the first structure to come out of the seed.

It goes downwards to produce the main root (primary root) of the plant from which subsequently come out many branch roots. In the meantime the stalks of the cotyledons elongate to make way for the plumule to come out, which actually goes upwards being pushed by the active epicotyl.

The epicotyl is the region between the plumule and the point of attachment of the cotyledons, i.e. the nodal zone. The plumule gradually grows upwards to produce the green aerial portions of the plant. Food matters, stored up in the cotyledons, are used up by the growing embryo, so the cotyledons gradually shrivel up.

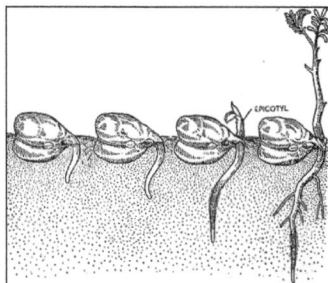

Stages in the germination Gram Seed (hypogeal).

It is to be noted that in the germination of gram seed the cotyledons remain inside the seed coat below the soil and the plumule goes upwards due to the active growth of the epicotyl. The position of the seed is not disturbed. This type of germination is called hypogeal or hypogenous (hypo=below; geo=earth). Peas and many other exalbuminous seeds show this type of germination.

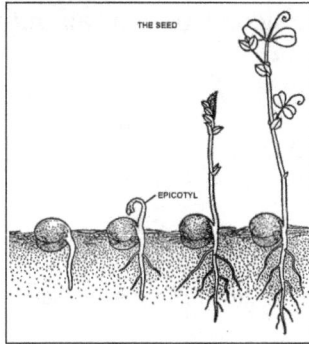

Stages in the germination Pea Seed (hypogeal).

3. Pea Seed

Pea seed is more or less rounded in shape. It has only one coat, the pale white testa. On the testa there is the distinct hilum or the scar, and the micropyle is present near about the hilum. On removing the seed coat the kernel is exposed.

It is made up of two fleshy cotyledons attached laterally to the axis of the embryo, the lower part of which is the radicle and the upper part, the plumule. So the kernel in pea corresponds to the embryo. Pea is also dicotyledonous exalbuminous.

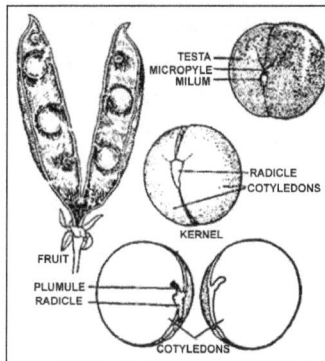

Structure of Pea Seed.

4. Castor-Oil Seed

It is more or less oblong in shape. The testa is hard, shell-like and has many sculpturing's on the outer surface. At one end the testa has a spongy outgrowth called the caruncle. The hilum and micropyle occur at that end and usually remain concealed by the outgrowth. Next to the hard brittle testa there is a thin white papery coat. It is not really the second coat or tegmen but is known as perisperm (remnants of nucellus).

The kernel consists of embryo and endosperm or albumen surrounding the embryo. The embryo has small radicle and very minute plumule forming the axis and two thin leafy cotyledons placed one against the other.

They have distinct veins which leave marking on the endosperm. Surrounding the embryo there is stored food matter, the endosperm. In a castor-oil seed the kernel is embryo plus albumen, and so it is a dicotyledonous albuminous seed.

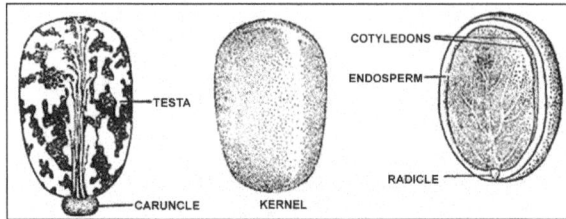

Structure of Castor-oil Seed.

- Dicotyledonous exalbuminous seeds: pea, gram, pulses, beans, mango

- Dicotyledonous albuminous seeds: castor-oil seed, papaw, poppy

5. Germination of Castor-Oil Seed

In the castor-oil seed the seed swells up, the testa bursts near the spongy outgrowth, the caruncle, and the radicle comes out first. It goes downwards as usual to produce the primary root which later on bears many branches. Now the hypocotyl, the region between the radicle and the nodal zone, grows fast forming a curved arch or loop.

This loop gradually straightens up, thus lifting the plumule and the cotyledons in the air above the soil level. Even then plumule and cotyledons remain surrounded by endosperm and remnants of test. The endosperm is gradually exhausted. The two cotyledons develop green colour and serve as the first pair of green leaves. Now plumule grows upwards to form the green aerial shoot.

So in the castor-oil seed the plumule goes upwards by the active growth and elongation of the hypocotyl which pushes the plumule, cotyledons, etc., above the soil level. Thus the position of the seed is definitely disturbed.

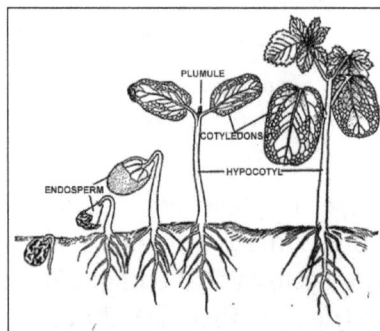

Stages in the germination of Castor-oil Seed (epigeal).

This is epigeal or epigeous germination (epi=upon, geo=earth). Gourd, tamarind also show this type of germination. In gourd the radicle usually develops, into the root and the Kypocotyl forms a loop which straightens up and pulls the rest above the surface of the soil.

A peg-like projection develops at the base of the hypocotyl which pushes the seed coat, so that the cotyledons and plumule may easily come out of the soil.

A peg like projection develops at the base of the hypocotyl which pushes the seed coat, so that the cotyledons and plumule may easily come out of the seed. The cotyledons become green and large and behave like ordinary leaves.

Tamarind seed also shows same type of germination. Here the testa is very hard and the two cotyledons are quite large and thick. They become greenish in colour but do not serve as green' leaves. The cotyledons gradually shrivel up and drop when the stored food matters are exhausted.

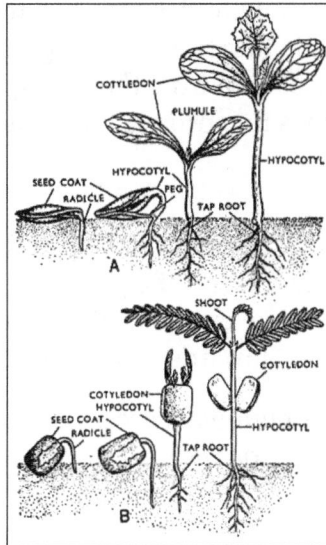

Stages in epigeal germination in gourd (A) and tamarind (B).

Monocotyledonous Seeds

1. Maize Grain

The maize grain is really a one-seeded fruit and not a seed, hence it is called a grain. The seed coat and the fruit wall or pericarp are inseparably united in the maize grain to form the outer pale yellowish coat.

A whitish deltoid area is noticed on one side of the grain which marks the position of the embryo. The embryo can be clearly seen if the grain is cut into two parts along the longer axis. It occupies only a very small portion of the grain, the remaining part being endosperm or stored food.

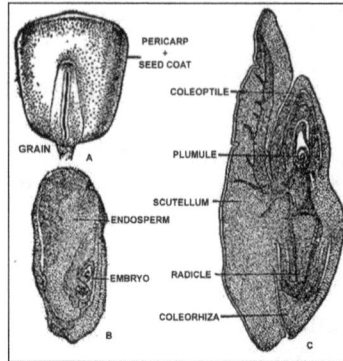

Structure of maize grain. A-grain; B-same cut lengthwise; C-embryo enlarged.

The embryo has an axis with radicle and plumule. The radicle is surrounded by a sheath called radicle sheath or coleorhiza, and the plumule has a similar sheath called plumule sheath or coleoptile.

A shield-shaped body, called scutellum, lies between the axis of the embryo and the endosperm, thus dividing the grain into two unequal parts. Scutellum is the single cotyledon of the maize grain. It absorbs food from the endosperm for the embryo to be used during germination. Maize grain is monocotyledonous albuminous.

2. Germination of Maize Grain

During germinations of the maize grain the radicle at first comes out boring through the coleorhiza or radicle sheath and forms the primary root. This root does not persist but dies and is replaced by a tuft of adventitious fibrous roots from the base of the stem. In the meantime the plumule goes upwards with the plumule sheath or coleoptile.

After growing to some extent the plumule pierces the sheath and grows up to produce the green aerial organs. Scutellum or the single cotyledon continues to supply food matters to the growing embryo during germination. The mode of germination in maize grain is also hypogeal. Rice and other grains also have this type of germination.

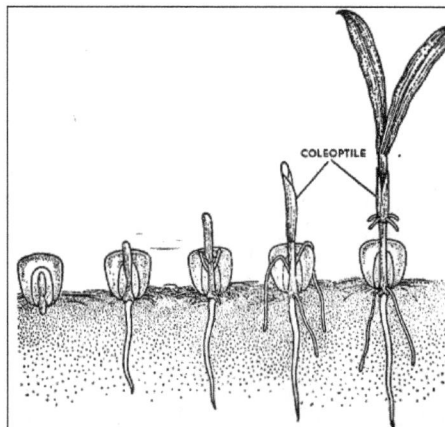

Stages in the germination of maize grain.

In cocoanut the innermost layer of fruit wall is the shell having three eye-like scars, and the embryo lies below one of the scars. The white kernel including the milk is the endosperm. During germination the lower end of the embryo—the cotyledon begins to grow as a spongy structure to absorb food materials stored in the endosperm.

The upper end of the embryo grows through the eye carrying the spongy structure to absorb food materials stored in the endosperm. The upper end of the embryo grows the eye carrying the radicle and the plumule; pierces through the fibrous coat and ultimately establishes itself.

Special Type of Germination

Plants growing in saline marshes or near sea-shore show a peculiar type of germination known as vivipary. In that case, the seeds begin germination before their liberation from the fruits. The radicle becomes elongated and considerably swollen.

Then the seed gets detached from the parent plant and comes vertically downwards. The radicle pierces the muddy soil below and thus gets fixed. Lateral roots are soon formed for anchorage, and the plumule is kept above the surface of Valine water. Examples: Rhizophora, Ceriops.

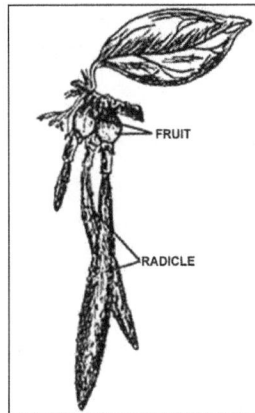

Viviparous germination.

Rice Grain

It is also a one-seeded fruit. The husk that encloses the grain is neither a part of the seed nor the fruit. It has four parts, two very small minute glumes, called empty glumes, at the base and two larger ones—the flowering glume and the palea, surrounding the grain.

The grain is exposed when the husk is removed. The reddish or whitish coat of the rice grain is the product of fusion of the pericarp and the seed coat. Here also the endosperm covers the major part of the grain, the minute embryo lying at one end beneath the palea.

During polishing of rice the outer coat and the embryo are removed, leaving a small break at one end of the grain. Obviously what remains after polishing is nothing but endosperm.

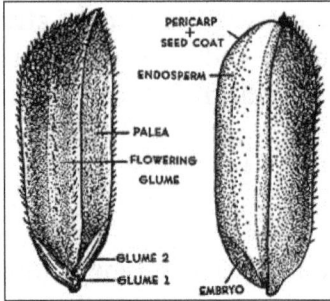

Structure of Rice Grain.

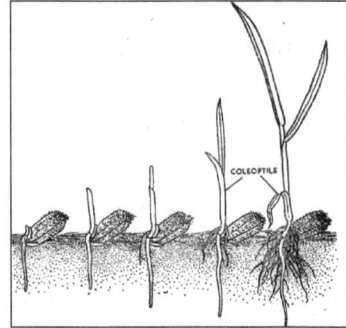

Stages in the germination of Rice Grain.

The embryo is more or less like that of maize having radicle and plumule with coleorhiza and coleoptile respectively and the shield-shaped scutellum, the single cotyledon. The radicle is a bit curved in rice grain. So it is also monocotyledonous albuminous. Most of the monocotyledonous grains are albuminous with a few exceptions like orchids.

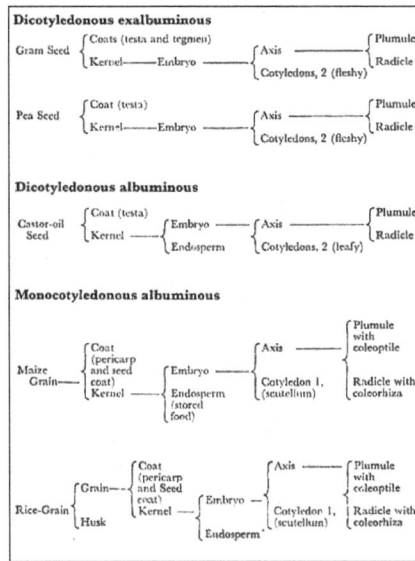

MORPHOLOGICAL FEATURES OF SEEDS

Morphology of a Gram Seed

Gram seed is a dicot, non-endospermic seed. The seeds are produced within the pods or leguminous fruits. A gram seed appears conical-pyriform in outline.

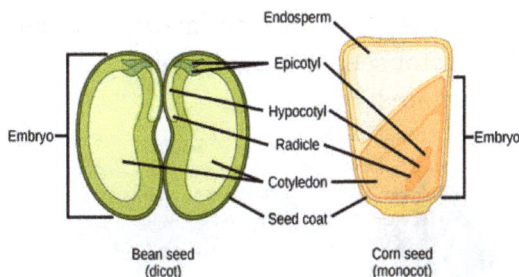

A. Structure of an endospermic seed, B. Structure of a non-endospermic seed.

It has following parts:

Seed Coat

It consists of two layers-outer testa and inner tegmen. Testa is thick and brownish. The tegmen is thin, membranous, and whitish and remains fused with testa. The pointed beak like end of the seed has a minute pore called micropyle. If a soaked seed is gently pressed, a drop of water oozes out of the micropyle. A small oval scar seen near the micropyle is called hilum through which the seed was attached to fruit. Another oval scar present in the middle is called chalaza or strophiole. A distinct ridge called raphe runs from hilum to chalaza.

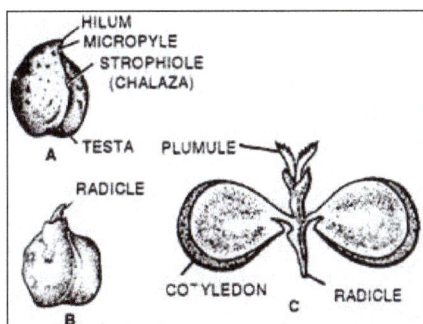

Gram seed: A. Whole seed, B. The kemel after remolval of seeds coat,
C. The two opened cotyledons with tigelium.

Embryo

It presents inner to seed coat. It consists of two circular yellowish cotyledons that are attached to the embryo axis. The part of embryo axis above the point of attachment to the cotyledons is called epicotyle. The tip of epicotyle is called plumule. Similarly, the region of the embryo axis below the point of attachment of cotyledons is called the hypocotyle. The tip of hypocotyle is called radicle. During germination, the radicle comes out first through the micropyle and grows to form a tap root. The plumule gives rise to shoot system.

Morphology of Castor Seed

Castor seed is a monocot, endospermic seed. The castor seeds are produced within a schizocarpic fruit called the regma which on maturity breaks up into 3 cocci, each containing a single seed. A castor seed is roughly oblong in outline with distinct convex (dorsal) and flat (ventral) surfaces.

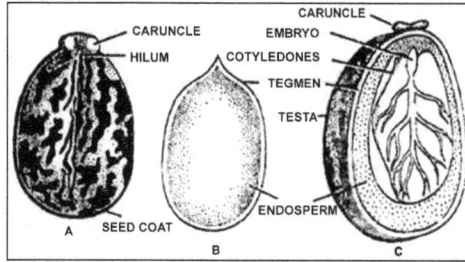

Castor seed: A. whole seed. B. kemel after removal of seed coat.
C. Whole seed cut into half showing position of embryc.

A castor seed has following parts:

Testa

It is the outer layer of seed coat. It is thick, hard and brittle. The external surface appears smooth, shinning and mottled brown in colour.

Tegmen

It is the inner layer of seed coat that appears dull and papery. Now it is called as perisperm or persistent nucellus.

Caruncle

It is a white spongy bilobed outgrowth present near the narrow end of the seed. If partially covers the hilum (dark scar) and completely covers the micropyle (small pore). Caruncle absorbs water which percolates through the micropyle into the seed.

Raphae

It is a shallow ridge present on the testa of flat surface of the seed. The distinct bifurcation of raphae represents chalaza.

Endosperm

It is a white oily food storage tissue that is present inner to the perisperm. From this layer castor oil of commerce is extracted.

Embryo

Embryo lies in the centre of endosperm. It consists of a radicle, a plumule and two lateral cotyledons, all of which are present on a short embryo axis. The cotyledons are thin, semi-transparent and oval in outline. They have palmate venation. The middle costa or rib is more prominent and bears a few lateral veins.

Radicle lies outside the cotyledons towards the micropylar end. It is a knob-like outgrowth. Plumule lies in between the two cotyledons and is quite indistinct. Epicotyl is also indistinct. In between the place of origin of the two cotyledons and the radicle is present a short hypocotyl. Castor-oil seed is dicotyledonous (having two cotyledons), endospermic (with a special food storing tissue called endosperm) and perispermic (having perisperm or persistent nucellus).

Morphology of a Maize Seed

Maize or Corn seed is actually a one seeded fruit called caryopsis or grain. It is a monocot endospermic seed.

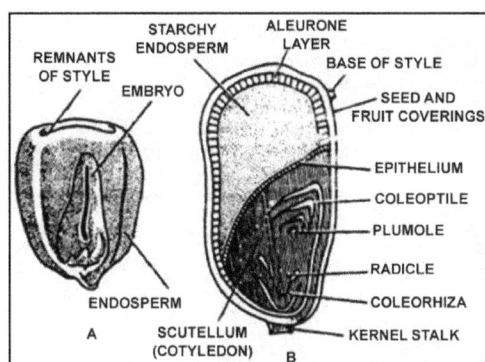

Monocotyledonous albuminous seed. A. Seed of Maize, B. L.S. of maize seed.

It consists of following parts:

Seed Coat

It is fused with the fruit wall (pericarp). It encloses a kernal which includes embryo and endosperm.

Endosperm

It constitutes 2/3 of the grain. Endosperm consists of outer aleurone layer and inner starchy endosperm.

Embryo

It lies on one side of the starchy endosperm and appears to be a lighter oval area in the whole seed. Embryo consists of a scutellum and a short embryo axis (tigellum). The scutellum is a shield-shaped cotyledon attached to a node of embryo axis. The surface of scutellum facing endosperm is called epithelial layer. It is both secretory and absorptive in nature. The epithelial layer secretes hormones into the endosperm for the synthesis of enzymes required for solubilisation of food. The solubilised food is absorbed by it and then transferred to the embryo axis.

The embryo axis has plumule (upper end) and radicle (lower end). The plumule contains a few rudimentary leaves and a conical protective sheath called coleoptile. The coleoptile has a termina pore for the emergence of first leaf during germination. The sheath is capable of growth. It assists the future shoot in passing through the soil during germination.

The radicle has two protective sheaths, inner root cap and outer coleorhiza. Roughly in the middle of embryo axis arises a vascular strand. It ramifies into the scutellum. The place of origin of the vascular strand from the embryo axis is called cotyledonary node.

SEED SCIENCE AND TECHNOLOGY

Seed science is the study of the structure and development of seeds from the moment of fertilization of the egg cell on the maternal plant until formation of a new plant from the seed. Seed science is the theoretical basis of seed growing. Agricultural seed science also elaborates methods of evaluating and controlling seed material. The science is closely connected with botany, biochemistry, genetics, and other biological sciences.

Seed technology are the methods through which the genetic and physical characteristics of seeds could be improved. It involves such activities as variety development, evaluation and release, seed production, processing, storage and certification.

"seed technology comprises techniques of seed production, seed processing, seed storage, seed testing and certification, seed marketing and distribution and the related research on these aspects".

Concept of Seed Technology

The distinction between seed and grain is vital, being of seminal importance to agriculture. A seed, strictly speaking, is an "embryo" a living organism embedded in the suppor ting or the food storage tissue. The seed pertains to material (seed, fruit or vegetatively propagating material) meant for saving for planting purposes, the essential function being the reproduction. The seed when scientifically produced (such as under seed certification) is distinctly superior in terms of seed quality, namely, the improved variety, varietal purity, freedom from admixtures of weeds and other crop seeds, seed health, high germination and vigour, seed treatment and safe moisture content etc. A grain on the other hand, includes cereals and pulses meant for human consumption.

Differences between Scientifically Produced Seed and The Grain (Used as seed).

S. No	(Scientifically produced) seed	Grain (used as seed)
1	It is the result of well planned seed programme	It is the part of commercial produce saved for sowing or planting purposes
2	It is the result of sound scientific knowledge, organized effort, investment on processing, storage and marketing facilities	No such knowledge or effort is required
3.	The pedigree of the seed is ensured. It can be related to the initial breeders seed	Its varietal purity is unknown
4.	During production, effort is made to rogue out off-types, diseased plants, objectionable weeds and other crop plants at appropriate stages of crop growth which ensures satisfactory seed purity and health	No such effort is made. Hence, the purity and health status may be inferior
5.	The seed is scientifically processed, treated and packed and labeled with proper lot identity	The grain used as seed may be manually cleaned. In some cases, prior to sowing it may also be treated. This is not labeled
6.	The seed is tested for planting quality namely, germination, purity, admixture of weed seeds and other crop seeds, seed health and seed moisture content	Routine seed testing is not done
7.	The seed quality is usually supervised by an agency not related with production (seed certification agency)	There is no quality control.
8.	The seed has to essentially meet the "quality standards". The quality is therefore well known. The labels, certification tags on the seed containers serves as quality marks.	No such standards apply here. The quality is non-descript and not known.

Role of Seed Technology

Feistritzer (1975) outlined the following roles of improved seed:

1. Improved seed: A carrier of new technologies.

The introduction of quality seeds of new varieties wisely combined with other inputs significantly increase yield levels. the cultivation of high yielding varieties have helped to increase food production from 52 million tonnes to nearly 180 million tonnes over a period of 40 years.

2. Improved seed: A basic tool for secured food supply.

The successful implementation of the high yielding varieties programme has led to a remarkable increase in production and food imports from other counters have been brought down inspite of rapid increase in population.

3. Improved seed: The principal means to secure crop yields in less favourable areas of production.

The supply of god quality seeds of improved varieties suitable to these areas is one of the important contribution to secure higher crop yields.

4. Improved seed: A medium for rapid rehabilation of agriculture in cases of natural disaster.

In case of floods and drought affected areas the Govt. will provide the improved seeds from national seed stocks to rehabilate the agricultural production of foods grains in the country.

Goals of Seed Technology

The major goal of seed technology is to increase agricultural production through the spread of good quality seeds of high yielding varieties. It aims at the following:

1. Rapid multiplication:

Increase in agricultural production through quickest possible spread of new varieties developed by the plant breeders. The time taken to make available the desired quantities of seeds of improved varieties to farmers should be considered as a measure of efficiency and adequacy in the development of seed technology in the country.

2. Timely supply:

The improved seeds of new varieties must be made available well in time, so that the planting schedule of farmer is not disturbed and they are able to use good seed for planting purposes.

3. Assured high quality of seeds:

This is necessary to obtain the expected dividends from the use of seeds of improved varieties.

4. Reasonable price:

The cost of high quality seed should be within reach of the average farmer.

Deterioration of Crop Varieties and Methods to Prevent them

1. Variety: is a group of plants having clear distinguished characters which when reproduced either sexually or asexually retains these characters.

The main aim of seed production is to produce genetically pure and good quality seed. But why/how the genetic purity of a variety is lost or deteriorated during seed multiplication. The several factors that are responsible for loss of genetic purity during seed production as listed are:

- Developmental variation

- Mechanical mixtures

- Mutations

- Natural crossing

- Genetic drift

- Minor genetic variation

- Selective influence of diseases

- Techniques of the breeder

- Breakdown of male sterility

- Improper/defective seed certification system

2. Developmental Variation: When a seed crop is grown in difficult environmental conditions such as different soil and fertility conditions, under saline or alkaline conditions or under different photo-periods or different elevations or different stress conditions for several consecutive generations the developmental variations may arise as differential growth response.

To avoid or minimize such developmental variations the variety should always be grown in adaptable area or in the area for which it has been released. If due to some reasons (for lack of isolation or to avoid soil born diseases) it is grown in non-adaptable areas it should be restricted to one or two seasons and the basic seed i.e. nucleus and breeder seed should be multiplied in adaptable areas.

3. Mechanical Mixtures: This is the major source of contamination of the variety during seed production. Mechanical mixtures may take place right from sowing to harvesting and processing in different ways such as:

a) Contamination through field – self sown seed or volunteer plants.

b) Seed drill – if same seed drill is used for sowing 2 or 3 varieties.

c) Carrying 2 different varieties adjacent to each other.

d) Growing 2 different varieties adjacent to each other.

e) Threshing floor.

f) Combine or threshers.

g) Bags or seed bins.

h) During seed processing.

To avoid this sort of mechanical contamination it would be necessary to rogue the seed fields at different stages of crop growth and to take utmost during seed production, harvesting, threshing, processing etc.

4. Mutations: It is not of much importance as the occurrence of spontaneous mutations is very low i.e. 10-7. If any visible mutations are observed they should be removed by rouging. In case of vegetatively propagated crops periodic increase of true to type stock would eliminate the mutants.

5. Natural Crossing: It is an important source of contamination in sexually propagated crops due to introgression of genes from unrelated stocks/genotypes. The extent of contamination depends upon the amount of natural cross-fertilization, which is due to natural crossing with undesirable types, offtypes, and diseased plants. On the other hand natural crossing is main source of contamination in cross-fertilized or often cross-fertilized crops. The extent of genetic contamina tion in seed fields is due to natural crossing depends on breeding system of the species, isolation distance, varietal mass and pollinating agent.

To overcome the problem of natural crossing isolation distance has to be maintained. Increase in isolation distance decreases the extent of contamination. The extent of contamination depends on the direction of the wind flow, number of insects presents and their activity.

6. Genetic drift: When seed is multiplied in large areas only small quantities of seed is taken and preserved for the next years sowing. Because of such sub-sampling all the genotypes will not be represented in the next generation and leads to change in genetic composition. This is called as genetic drift.

7. Minor Genetic variation: It is not of much importance, however some minor genetic changes may occur during production cycles due to difference in environment. Due to these changes the yields may be affected.

To avoid such minor genetic variations periodic testing of the varieties must be done from breeder's seed and nucleus seed in self-pollinated crops Minor genetic variation is a common feature in often cross-pollinated species; therefore care should be taken during maintenance of nucleus and breeder seed.

8. Selective influence of Disease: Proper plant protection measures much be taken against major pests and diseases other wise the plant as well as the seeds get infected.

   a. In case of foliar diseases the size of the seed gets affected due to poor supply of carbohydrates from infec ted photosynthetic tissue.

   b. In case of seed and soil borne diseases like downy mildew and ergot of Jowar, smut of bajra and bunt of wheat, it is dangerous to use seeds for commercial purpose once the crop gets infected.

c. New crop varieties may often become susceptible to new races of diseases are out of seed production programms. Eg. Surekha and Phalguna became susceptible to gall midge biotype 3.

9. Techniques of the Breeder: Instability may occur in a variety due to genetic irregulaeities if it is not properly assessed at the time of release. Premature release of a variety, which has been breed for particular disease, leads to the production of resistant and susceptible plants which may be an important cause of deterioration. When sonalika and kalyansona wheat varieties were released for commercial cultivation the genetic variability in both the varieties was still in flowing stage and several secondary selections were made by the breeders.

10. Breakdown of male sterility: Generally in hybrid seed production if there is any breakdown of male sterility in may lead to a mixture of F1 hybrids and selfers.

11. Improper Seed Certification: It is not a factor that deteriorates the crops varieties, but is there is any lacuna in any of the above factors and if it has not been checked it may lead to deterioration of crop varieties.

Maintenance of Genetic Purity during Seed Production

Horne (1953) had suggested the following methods for maintenance of genetic purity:

- Use of approved seed in seed multiplication.

- Inspection of seed fields prior to planting.

- Field inspection and approval of the Crop at critical stages for verification of genetic purity, detection of mixtures, weeds and seed borne diseases.

- Sampling and sealing of cleaned lots.

- Growing of samples with authentic stocks or Grow -out test.

Various steps suggested by Hartman and Kestar (1968) for maintaining genetic purity are as follows:

- Providing isolation to prevent cross fertilization or mechanical mixtures.

- Rouging of seed fields prior to planting.

- Periodic testing of varieties for genetic purity.

- Grow in adapted areas only to avoid genetic shifts in the variety.

- Certification of seed crops to maintain genetic purity and quality.

- Adopting generation system.

Safe guards for maintenance of genetic purity. The important safe guards for maintaining genetic purity during seed production are:

- Control of seed source

- Preceding crop requirement

- Isolation

- Rouging of seed fields

- Seed certification

- Grow out test

1. Control of Seed Source: The seed used should be of appropriate class from the approved source for raising a seed crop. There are four classes of seed from breeder seed, which are given and defined by Association of Official Seed Certification agency (AOSCA).

    a. Nucleus Seed: It is handful of seed maintained by concerned breeder for further multiplication. The nucleus seed will have all the characters that he breeder has placed in it and it is of highest genetic purity. The quantity of nucleus seed is in kilograms.

    b. Breeder Seed: It is produced by the concerned breeder or sponsoring institute or and which is used for producing foundation seed. It is of 100% genetic purity. The label/tag issued for B/s is golden yellow in color. The quality of breeder seed is assured by the monitoring team constituted by the govt.

    c. Foundation Seed: It is produced from breeder seed and maintained with specific genetic identity and purity. It is produced on govt. farms or by private seed producers. The quality of foundation seed is certified by certification agency. It has genetic purity of above 98%. The certification tag or label issued for F/s is white in color.

2. Preceding Crop requirement: This has been fixed to avoid contamination through volunteer plants and also the soil borne diseases.

3. Isolation: Isolation is required to avoid natural crossing with other undesirable types, off types in the fields and mechanical mixtures at the time of sowing, threshing, processing and contamination due to seed borne diseases from nearby fields. Protection from these sources of contamination is necessary for maintaining genetic purity and good quality of seed.

4. Rouging of Seed Fields: The existence of off type plants is another source of genetic contamination. Off type plants differing in their characteristics from that of the seed

crop are called as off types. Removal of off types is referred to as roughing. The main sources of off types are:

a. Segregation of plants for certain characters or mutations

b. Volunteer plants from previous crops or

c. Accidentally planted seeds of other variety

d. Diseased plants

Off type plants should be rouged out from the seed plots before they shed pollen and pollination occurs. To accomplish this regular supervision of trained personnel is required.

5. Seed Certification: Genetic purity in seed productions maintained through a system of seed certification. The main objective of seed certification is to make available seeds of good quality to farmers. To achieve this qualified and trained personnel from SCA carry out field inspections at appropriate stages of crop growth. They also make seed inspection by drawing samples from seed lots after processing. The SCA verifies for both filed and seed standards and the seed lot must confirm to get approval as certified seed.

6. Grow-out Test: varieties that are grown for seed production should be periodically tested for genetic purity by conducting GOT to make sure that they are being maintained in true form. GOT test is compulsory for hybrids produced by manual emasculation and pollination and for testing the purity of parental lines used in hybrid seed production.

SEEDLING

A seedling is a very young plant that grows from a seed. When the moisture, light, and temperature conditions are correct, the seedling's development begins with seed germination and the formation of three main parts:

1. Radicle - Embryonic root

2. Hypocotyl - Embryonic shoot

3. Cotyledons - Seed leaves

Seeds require the proper amount of light, temperature, oxygen, and water to germinate and form seedlings. After shooting, the seedling grows gradually while its food storage tissue shrinks. The seedling develops a root system and leaves to begin photosynthesis.

Monocots (one-seed leaves) and dicots (two-seed leaves) differ in early seedling development. In monocots, the primary root is protected by a coating known as a coleorhiza, which pushes off the seed first. Then the seedling leaves appear, also covered with a protective casing known as a coleoptile. On the other hand, in dicots a primary root (radicle) emerges first and anchors the seedling to the ground and permits it to start absorbing water. After absorbing water the shooting emerges. Unlike monocots, dicots emerge above the surface with their seed coat (epigeous germination).

Seedling Quality

Seedling quality has two main aspects. The first is the genetic quality or the source of the seed. The second component of seedling quality is its physical condition when it leaves the nursery. Improving genetic quality of seedlings requires a long term strategy of seed selection, while improving the physical quality can be accomplished in just one or two seasons.

Seed Source Quality

Farmers select only the best animals for breeding: animals that are small and sickly do not produce good offspring. Similarly, farmers use only the superior crops that have high yields and are resistant to disease for the next year's seed. These same principles should apply to trees. The characteristics of the parent trees can greatly influence the characteristics of the seedlings. The seed can determine whether the tree will grow well or poorly. Studies from around the world have shown that good seed improves survival, timber and fruit quality, and shortens rotation or harvest times. Because trees take longer to mature than crops or animals, thus making tree planting a long-term investment of labour and land, it is even more important to select only high quality seed.

The desired characteristics of the parent trees will vary depending on whether the trees are for wood, fodder, fruit, or medicine. A good nursery practice is to consult farmers as well as forestry technicians when selecting the seed sources. Farmers often know additional traits which make their trees more valuable. It may be difficult to find some of the trees with the best traits because these are often the first to be cut down. Conserving some of the best trees within the community will ensure a future supply of seed.

Some desirable parent tree characteristics are:

- Healthy trees with a large, well developed crown.

- For timber trees, a long, straight trunk with few branches.

- Wood quality, such as high density, or straightness of the grain.

- For fodder trees, palatability and digestibility of foliage (leaves that animals like to eat and are easily converted into energy).

- For fruit trees, low branching may be desired for easier fruit harvest.

- Fruit quality, such as sweetness or ability to be transported with minimum damage.

- Fast growth rate.

- Low susceptibility to (or ability to quickly recover from) disease or insect attack.

A good nursery practice is to select the parent trees well in advance, and plan a way to ensure sufficient seed is collected. Permanently marking the trees as seed sources may help ensure that they will not be cut down. Only mature seed from ripened fruit should be harvested. A good nursery practice for each species of tree is to collect seed from at least 30 parent trees that are at least 100 metres apart. If you buy the seed, find out how many trees were used. Using a large number of seed sources increases the gene pool or genetic diversity of the seedlings. Genes are the codes of information from the parent trees that determine how the progeny or offspring trees will grow. By using seed from many different trees, the probability of the offspring trees having good characteristics increases and ensures that the trees can better adapt to environmental changes. A poor, but unfortunately very common nursery practice is to collect seed from just one, two or three easy to climb trees close to the nursery. If the seed is bad and does not germinate, the nursery crop could be lost. Collecting from just a few trees is also dangerous because it results in low genetic diversity. Trees with low genetic diversity are often more susceptible as a group to disease, or unable to adapt to changing environmental conditions such as drought. If an area is planted with trees from very few sources, in the future, our ability to choose the best seed sources and improve the characteristics of the trees is very limited.

Use seed from an area as similar as possible to the area where you are planting. For example, seed from a mountainous region should only be planted in a mountainous region and seed originating from the lowlands will grow best in lowland conditions. If you purchase the seed, ask for its origin. It is okay to mix the seed from different trees together for normal nursery production. For long term genetic improvement though, seed from each individual tree is kept separately and tested in field experiments. The best trees are then selected to serve as seed sources for the nursery.

The principles of tree domestication are similar to those used in agriculture: maximize the quality of tree products, maximize tree growth rates, ensure the adaptability of species to the planting site, and maximize resistance to diseases and pests. This is achieved by selecting the best seed sources and managing the trees under optimal conditions.

Seedling Physical Quality

No single characteristic determines seedling quality. Seedling quality is a combination of height, diameter, plant nutrition, health, root size and shape. Together, these characteristics determine how well the plant will establish itself in the field, and they affect the rate of survival. Height alone is often not a good predictor of how a plant will grow in the field. A good nursery practice is to judge seedling quality by several traits.

Many of these traits act together and influence one another. The goal of producing the best seedling is to optimize these traits while producing specifically what is needed for a particular site. You will need to talk to foresters and farmers to find out the most important desired traits. For example, plants for dry, rocky soils may need to be short and be produced in small containers, whereas plants for flooded sites or active pastures may need to be quite large.

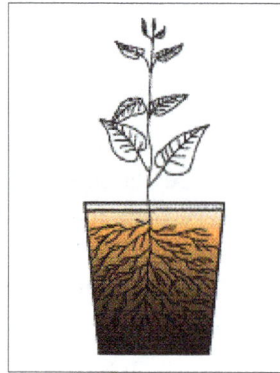

A quality seedling.

Quality tree seedlings have the following characteristics:

- They are healthy, vigorously growing and free of diseases.
- They have a robust and woody (lignified) single stem free of deformities.
- Their stem is sturdy and has a large root collar diameter.
- Their crown is symmetrical and dense.
- They have a root system that is free of deformities.
- They have a dense root system with many fine, fibrous hairs with white root tips.
- They have a 'balance' between shoot and root mass.
- Their leaves have a healthy, dark green colour.
- They are accustomed to short periods without water.
- They are accustomed to full sunlight.

The following examples demonstrate how these traits enable the plants to be more resistant to transport and planting stress, and how they improve seedling growth and survival.

SEED STORAGE PROTEINS

The plant seed is not only an organ of propagation and dispersal but also the major plant tissue harvested by humankind.

The amount of protein present in seeds varies from ~10% (in cereals) to ~40% (in certain legumes and oilseeds) of the dry weight, forming a major source of dietary protein. Although the vast majority of the individual proteins present in mature seeds have either metabolic or structural roles, all seeds also contain one or more groups of proteins that are present in high amounts and that serve to provide a store of amino acids for use during germination and seedling growth. These storage proteins are of particular importance because they determine not only the total protein content of the seed but also its quality for various end uses. For example, the low content of lysine, threonine, and tryptophan in various cereal seeds and of cysteine and methionine in legume seeds is due to the low proportions of these amino acids in the major storage proteins and may limit the nutritional quality of the seeds for monogastric animals. In the case of wheat, the storage proteins form the gluten fraction, whose properties are largely responsible for the ability to use wheat flour to make bread, other baked goods, and pasta. These properties are not shared by the storage proteins of other cereals.

Characteristics of Seed Storage Proteins

Despite wide variation in their detailed structures, all seed storage proteins have a number of common properties. First, they are synthesized at high levels in specific tissues and at certain stages of development. In fact, their synthesis is regulated by nutrition, and they act as a sink for surplus nitrogen. However, most also contain cysteine and methionine, and adequate sulphur is therefore also required for their synthesis. Many seeds contain separate groups of storage proteins, some of which are rich in sulphur amino acids and others of which are poor in them. The presence of these groups may allow the plant to maintain high levels of storage protein synthesis despite variations in sulphur availability. The strict tissue specificity of seed storage protein synthesis contrasts with that of tuber storage proteins, which may be synthesized in vegetative tissues under unusual conditions (for example, in vitro or after removal of tubers). A second common property of seed storage proteins is their presence in the mature seed in discrete deposits called protein bodies, whose origin has been the subject of some dispute and may in fact vary both between and within species. Finally, all storage protein fractions are mixtures of components that exhibit polymorphism both within single genotypes and among genotypes of the same species. This polymorphism arises from the presence of multigene families and, in some cases, proteolytic processing and glycosylation.

Classification of Seed Storage Proteins

Because of their abundance and economic importance, seed storage proteins were among the earliest of all proteins to be characterized. For example, wheat gluten was first isolated in 1745, and Brazil nut globulin was crystallized in 1859. However, the detailed study of seed storage proteins dates from the turn of the century, when Osborne classified them into groups on the basis of their extraction and solubility in water (albumins), dilute saline (globulins), alcoholhater mixtures (prolamins), and dilute acid or alkali (glutelins). The major seed storage proteins include albumins, globulins, and prolamins.

2S Albumin Storage Proteins

The 2s albumins were initially defined as a group on the basis of their sedimentation coefficients ($S_{20.w}$) of ~2. They are widely distributed in dicot seeds and have been most widely studied in the Cruciferae, notably oilseed rape (in which they are called napins) and Arabidopsis. The napins consist of two polypeptide chains with M_r values of ~9000 and 4000, which are linked by interchain disulfide bonds. They are synthesized as single precursor proteins that are proteolytically cleaved with the loss of a linker peptide and short peptides from both the N and C termini. This appears to be the most typical 2S albumin structure: similar heterodimeric proteins are present in species as diverse as pumpkin, cotton, castor bean, and lupin. The presence of two interchain bonds has been directly demonstrated in 2S albumins from lupin. Variant types of 2S albumin also occur. Those of pea appear to lack interchain disulfide bonds, whereas the 2S albumins of sunflower remain uncleaved. In addition, in sunflower and castor bean, some mRNAs encode two mature albumin proteins, each consisting of one or two subunits.

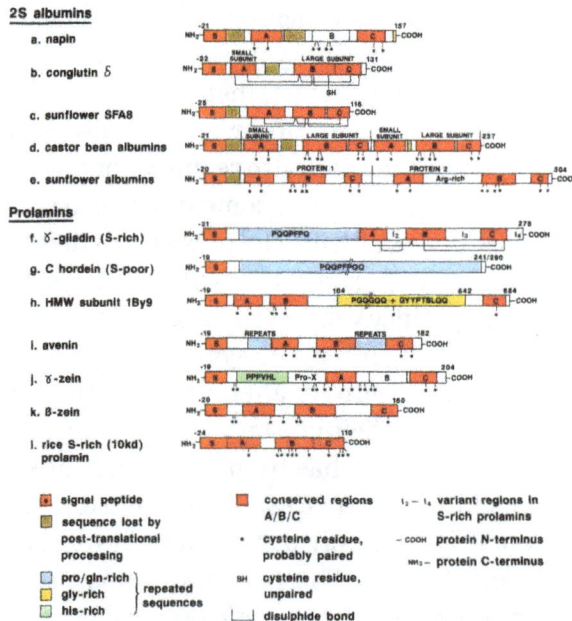

Schematic Structures of Members of the Cereal Prolamin Superfamily.

In figure, the cereal prolamin superfamily comprises the 2S albumins of dicots, the prolamins of the Triticeae, oats, and rice, and the αand β-zeins of maize. Three conserved regions (A, B, and C) are present in all except C hordein, although their boundaries are often poorly defined. These three regions also show homology with each other and contain cysteine residues that may be conserved within or between the different groups of proteins. For example, the 2S albumins shown all contain eight cysteine residues that are conserved in terms of context and position, including Cys-Cys and Cys-Xaa-Cys motifs, which are present in many of the other proteins.

Despite differences in their subunit structure and synthesis, all the 2S albumins are compact globular proteins with conserved cysteine residues. Although little is known about the detailed three-dimensional structures of 2S albumins, that of yellow mustard has been reported to contain ~50% a-helix, with little or no P-sheet. The authors proposed a ring structure with tightly packed a-helices, as suggested for zeins, but there is no experimental evidence for this structure.

Much of the recent interest in 2S albumins has focused on their exploitation in genetic engineering. Most notably, Altenbach et al. have used the 2S albumin of Brazil nut, which is rich in methionine, to increase the methionine content of tobacco seeds by up to 30%, and Higgins and co-workers have used the methioninerich sunflower 2S albumin SFA8 to increase the methionine content of forage grasses. In addition, the 2S albumins of Arabidopsis have been used as "hosts" for the synthesis of biologically active peptides, including the pentapeptide Leu-enkephalin and a 28-residue antibacterial peptide from Xenopus. In this work, the peptide was expressed as an insert within a variable loop region of the 2S albumin and then isolated by enzymatic cleavage. Although yields in oilseed rape equivalent to 1 kg of a25-residue peptide per hectare were achieved, the commercial viability of the work is uncertain.

Prolamin Storage Proteins

Whereas the 2s albumin and globulin storage proteins are widely distributed in flowering plants, the prolamins are restricted to one family, the grasses. These include the major cereals, in which prolamins usually account for approximately half of the total grain nitrogen. Exceptions to this general rule are oats and rice, in which the major storage proteins are 11s globulin-like and prolamins are present at low levels (~5 to 10Vo of the total grain protein).

Prolamins are traditionally recognized as a group on the basis of their solubility in alcoholhater mixtures (usually 60 to 70% [v/v] ethanol or 50% [v/v] propan-1-ol) and their high levels of glutamine and proline. However, comparisons of amino acid sequences have shown that this definition must be widened to include components that are insoluble in aqueous alcohols in the native state dueto the presence of interchain disulfide bonds and to recognize that all prolamins, even those that are insoluble in aqueous alcohols, are related, except for the a-zeins of maize (and their homologs present in related Panicoid cereals). All other prolamins form a single group known as the prolamin superfamily.

The Prolamin Superfamily of the Triticeae

Our understanding of this storage protein family stemmed initially from studies of the temperate cereals of the tribe Triticeae: barley, wheat, and rye. The prolamins of all three species are highly polymorphic mixtures of components whose M_r values range from ~30,000 to 90,000. These prolamins are classified into three groups -the S-rich, S-poor, and high molecular weight (HMW) prolamins-based on their amino acid sequences. Typical structures of the three types of prolamin are summarized in figures.

The S-rich prolamins are the quantitatively major prolamin group in all three species, accounting for ~80 to 90% of the total prolamin fractions. They include polymeric (that is, with interchain disulfide bonds) and monomeric (with intrachain disulfide bonds) components and consist of at least two families in each species: the 6 and γ-hordeins of barley; two types of γ-secalin of rye; and the a-gliadins, γ-gliadins, and low molecular weight (LMW) glutenin subunits of wheat. Their amino acid sequences consist of two separate domains: an N-terminal domain composed of repeated sequences, and a nonrepetitive C-terminal domain. The repetitive domain consists of tandem or interspersed repeats based on one or two short peptide motifs rich in proline and glutamine; this structure accounts for the high proportions of these two residues in the protein as a whole. For example, the repetitive domain of the γ-gliadin shown in figure is based on a Pro-Gln-GlnPro-Phe-Pro-Gln heptapeptide. This domain forms a secondary structure containing β-reverse turns and poly-L-proline II helix, as discussed later for the S-poor prolamins. In contrast, the nonrepetitive domain appears to have a globular structure rich in α-helix. This domain also contains most or all of the cysteine residues. Eight cysteines are present in the monomeric γ-gliadin, which form four intrachain disulfide bonds. Six of these cysteine residues are also present in the monomeric α-gliadins (based on sequence context); additional "unpaired" cysteine residues present in the polymeric LMW glutenin subunits may be responsible for polymer formation.

The S-poor prolamins include C hordein of barley, the ω-secalins of rye, and the ω-gliadins of wheat. Several genes encoding ω-secalins and C hordeins have been isolated. In all cases, the encoded proteins consist almost entirely of repeats of the octapeptide motif Pro-Gln-Gln-Pro-Phe-Pro-Gln-Gln that are flanked at the N-terminal side by short unique sequences of 12 residues and at the C-terminal side by short unique sequences of either six or four (ω-secalin) residues. The S-poor prolamins generally lack cysteine residues and therefore cannot form oligomers or polymers. Structural studies of C hordein indicate that the highly conserved repetitive primary structure results in a similarly conserved supersecondary structure. This is a loose spiral based on elements of β-turn and poly-L-proline II helix, the whole molecule forming a "stiff worm-like coil" of ~70 nm in length.

The HMW prolamins are typified by the HMW subunits of wheat glutenin, which have been studied in detail because of their putative role in determining the elasticity, and hence the bread-making performance, of wheat doughs. Extensive repeated sequences are present, flanked by nonrepetitive N- and C-terminal domains. The repeated sequences are based on the motifs Gly-Tyr-Tyr-Pro-Thr-Ser-Pro or Leu-Gln-Gln, Pro-Gly-Gln-Gly-Gln-Gln, and, in some subunits only, Gly-Gln-Gln. Differences in the number of repeated peptides are largely responsible for variation in HMW subunit size.

Although the repeated sequences present in the HMW subunits are not related to those in the S-poor prolamins, they appear to adopt a similar spiral supersecondary structure, although one that is more compact because it includes p-turns but not poly-L-proline II structure. The net result is a rod-shaped molecule, which has been imaged directly by scanning probe microscopy. As in the S-poor prolamins, cysteine residues are largely restricted to the nonrepetitive domains. These domains appear to be globular (being rich in α-helix), with the cysteine residues allowing the formation of an elastic network stabilized by interchain disulfide bonds.

Evolutionary Relationships among the S-Rich, S-Poor and HMW Prolamins

The three groups of prolamins present in the Triticeae all consist of at least two discrete domains, one of which is based on repeated sequences. More detailed comparisons show that the prolamins are likely to have evolved from a single ancestral protein. Comparisons of the nonrepetitive domains of a range of S-rich prolamins show that all contain three conserved regions of between 20 and 30 residues. These regions, designated A, B, and C, contain most of the conserved cysteine residues and are also related to each other, indicating that they are likely to have originated from the triplication of a short ancestral domain. Insertion of additional variable regions (1, to 14) and of repeated sequences at the N-terminal side of region 1, would have given rise to the range of presentday S-rich prolamins. Short regions related to A, B, and C is also present in the HMW prolamins, although in this case regions A and B are in the N-terminal domain and region C is in the C-terminal domain. Therefore, these proteins are likely to have evolved from the same ancestor as did the S-rich prolamins, although unrelated repeated sequences have been inserted between regions B and C.

The S-poor prolamins are also clearly related to the S-rich prolamins in that their repetitive sequences are based on similar proline- and glutamine-rich peptide motifs. For example, the heptapeptide and octapeptide motifs present in γ-gliadin and C hordein differ in only a single glutamine residue. The S-poor group is hypothesized to have evolved from the S-rich prolamins by further amplification of the repeated sequences and deletion of most of the non-repetitive domain that contains regions A, B, and C.

The Prolamin Superfamily in other Species

Prolamins related to those present in the Triticeae are also present in a range of other cereals. These include oats, in which the avenins contain regions A, B, and C together with two blocks of repeats rich in proline and glutamine, and rice. The prolamins of rice consist of three groups of small proteins. Although these do not contain repeated sequences, they appear to be related to one another and to the prolamins of the Triticeae. For example, the sulfurrich M_r, 10,000 prolamins shown in figure appear to contain vestiges of regions A, B, and C.

The prolamins of maize, known as the zeins, and of related Panicoid cereals such as sorghum, pearl millet, and Job's tears, fall into four groups, three of which belong to the prolamin superfamily. In maize, these are the β-, γ-, and δ-zeins. The β-zeins and γ-zeins both contain regions related to A, B, and C. The δ-zeins do not contain repeats or any other distinguishing features, but homology with the prolamin superfamily can be inferred from some sequence identity with the 2S albumin of Brazil nut. All three of these groups of zeins are rich in cysteine and methionine, residues deficient in the α-zeins.

The 2S Albumins are also Related to the Prolamin Superfamily

The 2S albumins also contain three conserved regions related to regions A, B, and C. These regions contain the eight conserved cysteine residues present in most 2S albumins, with region A and regions B and C corresponding to the small and large subunits, respectively, of the heterodimeric 2S albumins. The absence of repeated sequences and the widespread distribution of 2S albumins in dicots may indicate that they are similar to the ancestral protein of the prolamin superfamily, although this would have lacked the proteolysis site between regions A and B.

The α-Zeins of Maize

The α-zeins account for ~75 to 80% of the total prolamins in maize and are classified into two groups with slightly different M_r (~19,000 and ~22,000). They have similar structures, consisting of unique N- and C-terminal domains flanking repeated sequences. Although the latter are generally considered containing blocks of ~20 residues, they are highly degenerate, with no clear consensus motif. There is no evidence of homology with the repeated sequences present in other prolamins, and the unique N- and C-terminal sequences do not appear to be related to any other protein. The size difference between the M_r, 19,000 and M_r, 22,000 zeins may result from variation in the number of blocks present in the repetitive domains (nine and 10, respectively) or from the insertion of a loop region of ~20 residues in the C-terminal domain of the M_r, 22,000 proteins.

The precise structure adopted by the α-zeins is still uncertain. Whereas a range of biophysical studies demonstrates that they have extended conformations when in solution, they may adopt a more compact conformation when present in the hydrated solid state, that is, in protein bodies. For example, Argos et al. proposed that α-zeins form an antiparallel ring of nine α-helices, facilitating packaging in the protein bodies.

Globulin Storage Proteins

The globulins are the most widely distributed group of storage proteins; they are present not only in dicots but also in monocots (including cereals and palms) and fern spores. They can be divided into two groups based on their sedimentation coefficients ($S_{20,w}$): the 7S vicilin-type globulins and the 11S legumin-type globulins. Both groups show considerable variation in their structures, which results partly from post-translational processing. In addition, both have nutritional significance in that they are deficient in cysteine and methionine, although 11S globulins generally contain slightly higher levels of these amino acids. The globulin storage proteins have been studied in most detail in legumes, notably peas, soybean, broad bean (Vicia faba), and French bean (Phaseolus vulgaris).

The 11S Globulins

The 11S legumins are the major storage proteins not only in most legumes but also in many other dicots (for example, brassicas, composits, and cucurbits) and some cereals (oats and rice). The mature proteins consist of six subunit pairs that interact non-covalently. Each of these subunit pairs consists in turn of an acidic subunit of M_r, ~40,000 and a basic subunit of M_r, ~20,000, linked by a single disulfide bond. Each subunit pair is synthesized as a precursor protein that is proteolytically cleaved after disulfide bond formation. Legumins are not usually glycosylated, an exception being the 12S globulin of lupin. This contrasts with the 7S globulins.

Although the 11S globulin of Brazil nut was one of the first proteins to be crystallized, the crystals of this and other 11S globulins have generally been small and disordered and have failed to provide any details of protein structure. However, a recent study of edestin, an 11S globulin from hempseed, is more promising. Although the crystals showed some disorder, they exhibited enough symmetry so that some measurements could be made. These indicated that the subunits are arranged in an open ring structure, oriented alternately up and down, in a disk whose diameter is 145 Å and whose thickness is ~90 Å.

The 7S Globulins

7S globulins are typically trimeric proteins of M_r, ~150,000 to 190,000 that lack cysteine residues and hence cannot form disulfide bonds. Their detailed subunit compositions vary considerably, mainly because of differences in the extent of post-translational processing (proteolysis and glycosylation). For example, the vicilin subunits of pea are initially synthesized as groups of polypeptides of M_r, ~47,000 and ~50,000, but post-translational proteolysis and glycosylation then give rise to subunits with M_r, values between 12,500 and 33,000. These subunits are difficult to purify and characterize, but molecular cloning allowed their origins and the sites of proteolytic cleavage and glycosylation to be identified.

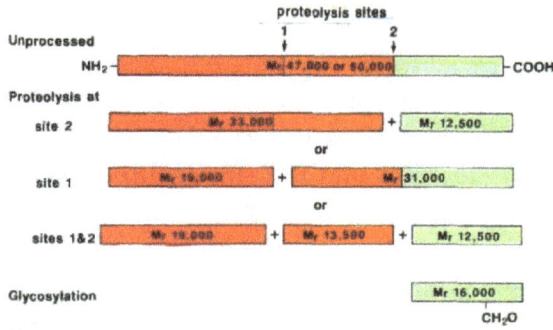

Schematic Diagram Showing the Origin of the Pea Vicilin Subunits.

Based on Gatehouse, the M_r 19,000, M_r 13,500, and M_r 12,500/16,000 subunits are shown in red, orange, and green, respectively.

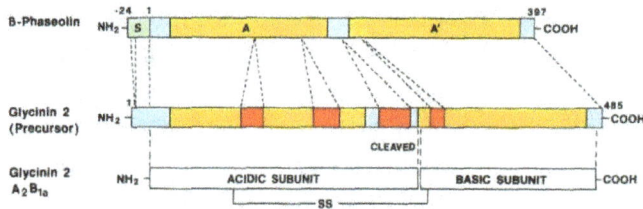

Alignment of Phaseolin and Glycinin 2 Subunit Sequences.

In figure, the alignment of β-phaseolin (French bean 7S globulin) and glycinin 2 (soybean 11S globulin) is based on the structure of phaseolin. Two structurally similar units, A and A' (shown in yellow), have been defined by x-ray crystallography. Each unit consists of a β-barrel with a "jelly roll" motif followed by an α-helical domain comprising three helices. Unit A is located in the acidic subunit of the 11S protein and unit A' in the basic subunit. The blue areas flanking and separating regions A and A' correspond to regions of sequence homology that are not reflected in the secondary structures as determined by crystallography. The distribution of globally conserved residues in the 7S/11S alignment indicates the presence of four major insertions in the 11S protein (shown in red). Two of these are in unit A and one is in unit A', all falling within loop regions connecting structural elements. The fourth insertion is in the region of sequence homology between A and A', at the C-terminal end of the acidic subunit. Insertions and deletions of less than six residues are not shown. The signal peptide of the p-phaseolin, shown in green, has limited sequence homology with the N-terminal region of the mature glycinin 2 precursor.

The 7S globulins of P. vulgaris and soybean differ from those of pea and V. faba in that glycosylation is more extensive but proteolysis does not occur. For example, the 7S phaseolin of P. vulgaris consists of glycosylated subunits with M_r, values between ~43,000 and 53,000.

The three-dimensional structures of several 7S globulins have recently been determined using x-ray crystallography. These show that the trimeric proteins are disk shaped, with diameters of ~90 Å and thicknesses of 30 to 40 Å.

11S and 7S Globulins are Related

Although the 7S and 11S globulins show no obvious sequence similarities, they do have similar properties, including the ability to form both trimeric and hexameric structures. In the case of the 7S globulins, the mature protein is trimeric, but it may undergo reversible aggregation into hexamers, depending on the ionic strength. The mature 11S globulin, by contrast, is hexameric but is initially assembled and transported through the secretory system as an intermediate trimer. Therefore, it is not surprising that more sophisticated comparisons have shown that the 11S and 7S globulin subunits are related in structure. Such comparisons indicate that the basic (C-terminal) chain of the 11S legumins is related to the C-terminal region of the 7S vicilins. Lawrence et al. determined the x-ray crystal structure of 7S phaseolin and established sequence homologies between 7S/11S proteins. The homologies showed that the 11S sequences, with four major insertions of sequence, can be aligned with 7S sequences. The authors further proposed that the 11S legumins have a tertiary structure similar to that of the 7S vicilins and concluded that the 7S and 11S proteins evolved from a common ancestral protein.

Synthesis, Assembly and Deposition of Seed Storage Proteins

All of the seed storage proteins discussed here are secretory proteins synthesized with a signal peptide that is cleaved as the protein is translocated into the lumen of the endoplasmic reticulum (ER). The subsequent events in storage protein processing are less clearly understood and may vary not only between different species but also within the same species, depending on the protein type and stage of development.

Storage Protein Folding and Assembly in the ER

Secretory proteins assume their folded conformations within the lumen of the ER, which is also the site of disulfide bond formation. Studies of other systems demonstrate that three types of ER lumenal proteins may assist in these processes. Molecular chaperones of the HSP70/BiP family may facilitate folding by binding transiently to the nascent polypeptides and may also prevent the formation of incorrect inter- or intramolecular interactions. BiP-related proteins are present in developing endosperms of cereals such as rice, wheat, and maize, and they accumulate in higher than normal levels in high-lysine maize mutants, possibly due to the presence of incorrectly folded zeins.

A second group of proteins, the peptidyl-prolyl cis-trans isomerases (PPI), or cyclophilins, of which one subclass (the S-cyclophilins) is resident in the ER lumen, may also assist folding by accelerating the isomerization of Xaa-Pro bonds, which is a rate-limiting step in the folding of some proteins. The repetitive domains of cereal prolamins contain high levels of proline, and isomerization of Xaa-Pro bonds might therefore be expected to limit their folding. This does not; however, appear to be the case, at least in vitro. For example, C hordein contains ~30 mol % proline residues, all of which appear to be in the Trans configuration. Nevertheless, it folds readily in vitro. Our preliminary studies also show that the levels of cyclophilinrelated transcripts decrease during the period of gluten protein synthesis in developing wheat seeds. 7S and 11S globulin subunits are also assembled in the ER, with the 7S globulins forming the mature trimers catalyzes disulfide bond formation in storage proteins also remains to be established, although Bulleid and Freedman showed that depletion of PDI from dog pancreas microsomes resulted in defective synthesis of disulfide bonds in a γ-gliadin synthesized in vitro. PDI has also been shown to be associated with the ER in developing wheat endosperms, although the levels of PDI transcripts peak somewhat earlier than those of gluten proteins. The assembly of some prolamins into disulphide-stabilized polymers presumably also takes place in the ER, although there is no information available on how this occurs. N-linked glycosylation of the 7S phaseolin subunits also occurs in the ER lumen, probably as a cotranslational event 1991.

Compartment	Event	2S Albumins	Prolamins	7S Globulins	11S Globulins
ER	Co-translational insertion	+[a]	+	+	+
	Signal peptide cleavage	+	+	+	+
	Chaperone-mediated folding (BiP)[b]	NA	NA	NA	NA
	S-S bond formation (PDI)	NA	+[c]	-	+
	Isomeration of Xaa-Pro bonds (PPI)[d]	NA	NA	NA	NA
	N-Glycosylation	-	-	+	-
Golgi	Complex glycan addition	-	-	+	-
Vacuole	Propeptide processing (cleavage at asparagine residues by proteases)	+[e]	-	+	+[f]

Where,

a. + , the process has been experimentally observed; - , the process has been looked for but not detected; NA, the process has not been examined.

b. Although BiP is likely to interact transiently with every elongating nascent chain translocating across the ER membrane, its subsequent role in folding is likely to be protein dependent.

c. Because the majority of storage proteins form disulfide bonds, it is assumed that this reaction is catalyzed by protein disulfide isomerase (PDI) in the ER lumen.

d. Peptidyl-prolyl cis-trans isomerase (PPl) may be required for the folding in the ER of proline-rich proteins such as the prolamins.

e. Aspartic and thiol proteases have been characterized from B. napus and castor bean, respectively.

f. Thiol proteases have been characterized from soybean and castor bean.

Whether protein disulfide isomerase (PDI) catalyzes disulfide bond formation in storage proteins also remains to be established, although Bulleid and Freedman showed that depletion of PDI from dog pancreas microsomes resulted in defective synthesis of disulfide bonds in a y-gliadin synthesized in vitro. PDI has also been shown to be associated with the ER in developing wheat endosperms, although the levels of PDI transcripts peak somewhat earlier than those of gluten proteins. The assembly of some prolamins into disulfide-stabilized polymers presumably also takes place in the ER, although there is no information available on how this occurs.

N-linked glycosylation of the 7S phaseolin subunits also occurs in the ER lumen, probably as a cotranslational event. Phaseolin has two consensus N-glycosylation sites, one of which is always used, whereas the other, located closer to the C terminus, is used less frequently. Wild-type phaseolin assembles into trimers in the ER, but trimerization is prevented if a C-terminal sequence of 59-amino acid residues is deleted. In this case, the protein monomer remains in the ER and becomes glycosylated at the second site. Moreover, this assembly-defective protein interacts with BiP in an ATPdependent manner, highlighting the role of BiP in binding to malfolded proteins.

Storage Protein Transport and Protein Body Formation in Cereals

Two routes of protein body formation appear to operate in developing cereal endosperms, in one of which the protein body forms from the vacuole and in the other of which it forms from the ER. For example, the major storage proteins in oats and rice are related to the 11S globulins of dicots and appear to be transported from the ER lumen via the Golgi apparatus to the vacuole. The protein bodies then appear to form by fragmentation of the vacuole. In contrast, the prolamins of rice and maize appear to be retained within the lumen of the ER, which becomes distended to form protein bodies. Thus, rice endosperm cells contain two populations of protein bodies, some of vacuolar origin (containing glutenins) and others of ER origin (containing prolamins).

The situation appears to be more complicated in barley, wheat, and rye, with prolamins present in both ER-derived and vacuolar protein bodies. In the case of wheat, these may differ in their protein content, as Rubin et al. suggested when they reported that glutenins are retained predominantly in ER-derived protein bodies, whereas gliadins are present in both types of protein body. In addition, Levanony et al. proposed that ER-derived protein bodies may subsequently fuse with the vacuoles, bypassing the Golgi apparatus.

The mechanisms that determine whether a prolamin is transported to the vacuole or retained in the ER are not known, because neither vacuolar targeting nor ER retention sequences have been identified. Li et al. have suggested that rice prolamins are retained in the ER by interaction with BiP, which itself has a C-terminal ER retention sequence. Although the work of Li et al. implies a "once only" binding of BiP to the emerging nascent polypeptide chain, it is now known that BiP binds to such chains in a sequential manner, "pulling" the protein into the ER lumen. This indicates that a stable interaction of BiP with a nascent prolamin chain is very unlikely. In fact, the only clear examples of BiP binding to storage proteins are to malfolded or assemblydefective forms. The expression of BiP-related transcripts is not coordinated with prolamin gene expression in developing endosperms of wheat, and expression of wheat y-gliadin in seed of transgenic tobacco plants does not alter the level of BiP transcripts, which suggests that BiP is unlikely to be involved in prolamin retention in the ER. Studies of y-gliadin transport and retention in Xenopus oocytes seem to indicate that prolamin accumulation in the ER does not require any plant-specific factors. Whereas a truncated form of the protein corresponding to the N-terminal domain accumulated in the ER to form protein body-like structures, a truncated form containing the C-terminal domain was secreted. The intact wild-type protein was also secreted, but at a lower rate than was the C-terminal domain. Although the prolamin repetitive domains could be responsible for ER retention by interacting with ER components, a simpler model is that interactions between the individual prolamin molecules result in the formation of insoluble aggregates that are not readily transported from the ER lumen. Such a model is supported by the observation of Li et al. that rice prolamin mRNAs are segregated to a distinct region of the rough ER. Such segregation could allow aggregation of the prolamins to occur in localized parts of the ER, preventing widespread effects on ER integrity.

Storage Protein Transport and Protein Body Formation in Dicots

The 2S albumins and 7S/11S globulins of legumes and other dicots are transported via the Golgi apparatus to the vacuole, which fragments to form protein bodies. Despite several attempts, specific vacuolar targeting sequences have not been identified in these proteins. Instead, it is probable that one or more exposed regions of the correctly folded protein are recognized by the sorting machinery.

The assembly of the 11S globulins appears to be a highly regulated event. The monomeric proteins are initially assembled in the lumen of the ER into trimers that are then transported from the ER to the storage vacuole, where they are assembled into their final hexameric form. This assembly process requires specific proteolytic cleavage of the subunits present in the trimers. Uncleaved trimers cannot assemble into hexamers in vitro unless they have been treated with papain. This cleavage does not cause the trimers to disassemble but may result in a conformational change that favors assembly into hexamers. The 11S globulin vacuolar processing protease has been characterized from several species and shown to recognize asparagine processing sites specifically. Scott et al. purified a soybean protease that cleaves the trimeric 11S globulin proproteins. Hara-Nishimuraet al. also purified a 11S globulin processing peptidase from castor bean that displays similar processing specificity and also appears to be a thiol protease but is unglycosylated.

The 11S globulin processing peptidase of Hara-Nishimura et al. is also able to cleave 2S albumin precursors in vitro at their asparagine processing sites. A 2S albumin processing protease that cleaves 2S albumin proproteins from Arabidopsis in vitro has also been characterized. Although this enzyme has the same specificity as that of the one purified by Hara-Nishimura et al., it is an aspartic protease rather than a thiol protease. The processing of 2S albumins does not appear to be required for their assembly, in contrast with the case of the 11S globulins.

Storage Protein Packaging

Little is known about how storage proteins are organized within protein bodies, although this organization may well be important in ensuring efficient use of storage space and facilitating mobilization of storage proteins during germination. Whereas prolamin and globulin storage proteins are present in separate protein bodies in rice, they are located within the same protein bodies (although in separate phases of them) in other cereals. Prolamin inclusions are present in a globulin matrix in oats, and globulin (triticin) inclusions are present within a prolamin matrix in wheat.

In leguminous plants, the 7S and 11S globulins appear to be in the same protein bodies with no spatial separation. As discussed previously, their structures may facilitate efficient packaging. In many other dicots, such as pumpkin, sunflower, brassicas, and castor bean, 2S albumins are stored together with 11S globulins, but how these distinct types of storage protein are organized in the protein bodies is not known. In contrast, there is evidence that the different types of prolamins are spatially separated in the protein bodies of cereal endosperms. This is most clear in maize, where immunogold labeling has shown that the α-zeins form the protein body core, with β- and γ-zeins at the periphery. The α-zeins may also be present in the protein body core. The evidence for spatial separation of prolamins in protein bodies of the Triticeae is less convincing, but Rechinger et al. proposed that the quantitatively minor S-rich$_{\gamma 1}$- and S-rich$_{\gamma 2}$-hordeins form a peripheral layer surrounding a core of B hordeins (S-rich) and C hordeins (S-poor) in barley. The separation of different proteins in cereal protein bodies could result from the properties of the proteins themselves (for example, their ability to separate into separate phases) or from different patterns of deposition during protein body biogenesis.

SEED TREATMENT

Seed treatments are defined as chemical or biological substances that are applied to seeds or vegetative propagation materials to control disease organisms, insects, or other pests.

Seed treatment pesticides include bactericides, fungicides, and insecticides. Most seed treatments are applied to true seeds, such as corn, wheat, or soybean, which have a seed coat surrounding an embryo. However, some seed treatments can be applied to vegetative propagation materials, such as bulbs, corms, or tubers (such as potato seed pieces).

Seed-applied growth regulators, micronutrients, and nitrogen-fixing Rhizobium and Bradyrhizobium inoculants are not included because they are not intended for pest control. Treatments designed to protect stored food or feed grain are considered grain treatments rather than seed treatments. Pest control in stored grain and storage facilities requires additional licensing.

Benefits and Risks

Seed treatments are used on many crops to control a variety of pests. Seed treatments are commonly used to ensure uniform stand establishment by protecting against soilborne pathogens and insects. In fact, they are considered so essential for corn stand establishment that virtually all corn seed is treated. Seed treatments have had phenomenal success in eradicating seedborne pathogens, such as smut or bunt, from wheat, barley, and oats. Seed treatments can be used to suppress root rots in certain crops. Finally, some newer systemic seed treatments can supplement or may provide an alternative to traditional broadcast sprays of foliar fungicides or insecticides for certain early-season foliar diseases and insects.

Although seed treatments have important benefits, they also pose certain risks. One risk is accidental exposure of workers who produce or apply seed treatments. Another risk is contamination of the food supply by accidental mixing of treated seed with food or feed grain. A third risk is accidental contamination of the environment through improper handling of treated seeds or seed treatment chemicals. All of these risks can be minimized by proper training and proper use of seed treatment pesticides.

Factors that Favor the use of Seed Treatments

- Field is for seed production. ï Low test weight or older seed.

- Planting in unfavorable germination conditions, such as dry soil or cold soil.

- Planting into fields with a history of stand establishment problems.

- Planting to precise populations.

- Replanting will not be feasible if first planting fails.

- Seed is expensive.

- Seed thought to carry certain seedborne pathogens.

- Yield potential of field is high.

Integrated Pest Management

Seed treatments should be considered as tools in an integrated pest management (IPM) plan. IPM is the use of a combination of cultural practices, host resistance, biological control, and chemical control methods to simultaneously (1) minimize economic losses due to pests, (2) avoid development of new pest biotypes that overcome pesticides or host resistance, (3) minimize negative effects on the environment, and (4) avoid pesticide residues in the food supply. An IPM plan should identify important pests, determine pest management options, and blend together various management options to achieve the goals.

To use seed treatments effectively, it is important to understand the purposes of seed treatment, alternatives or supplements to seed treatments, and the various advantages and disadvantages of seed treatments.

Purposes of Seed Treatment

Control of Seedborne Pathogens

Seedborne, disease-causing pathogens may occur on the surface of seed, hidden in cracks or crevices of seed, or as infections deep inside the intact seed. These pathogens may be important for three reasons. First, some pathogens do not survive in soil or crop residue and are dependent on the seedborne phase for survival between crops. An example is the fungus that causes loose smut of wheat. Second, even if a pathogen can survive in soil or residue, being seedborne may allow it to get a head start and, thus, result in more severe disease. An example would be the fungus that causes Septoria leaf blotch of wheat. Third, seedborne pathogens may hitch a ride to new localities in seed shipments (such as the fungus that causes Karnal bunt of wheat or the bacterium that causes black rot of crucifers).

Seed treatments can often be used to control pathogens that occur on or in the seed. The choice of seed treatment may be dictated by whether the pathogen is borne externally or internally. For example, both systemic and nonsystemic (contact) fungicides can eliminate surface contamination of wheat seed by spores of the common bunt fungus. However, the fungus causing loose smut of wheat is borne within the seed embryo and cannot be controlled with a contact fungicide. In that case, a systemic fungicide is required to control the internal pathogen.

Protection of Seeds and Seedlings

Seeds and seedlings are vulnerable to many soilborne and foliar pests. Insects and pathogens can destroy germinating seeds and young plants, which are relatively tender and lack food reserves to recover from injuries or to survive extended periods of stress. Examples of stress include heavy rains, crusted soils, compaction, deep planting, cool soil, very dry soils, and some postemergence herbicides. Under stressful conditions, a number of aggressive or even fairly weak pathogens can become active and cause plant population and yield losses.

Seed treatments can protect the seed and seedling from attack by certain insects and pathogens. Nonsystemic fungicides or insecticides form a chemical barrier over the surface of the germinating seed. This barrier protects the germinating seed from chewing insects, such as wireworms, or soilborne pathogens, such as pythium. Certain systemic seed treatments can protect aboveground parts from sucking insects, such as aphids, or foliage diseases, such as rust. Systemic fungicides and biological seed treatments can also protect young plants from root rot. Although the duration of protection may be limited, a delay in infection can reduce the losses. For chronic diseases, such as root rots, the earlier that the infection takes place, the greater will be the damage.

Typically, seed treatments will last only about 10 to 14 days beyond planting, with pesticide breakdown being most rapid under warm, moist conditions. However, certain active ingredients can protect seedlings considerably longer when applied at the highest labeled rate.

Alternatives or Supplements to Seed Treatment

Usually, seed treatments are not the only available method to control a particular pest. Seed treatments should be compared to alternative pest control measures for cost, efficacy, safety, and so on. Often, no single pest control method provides sufficient control. Seed treatments can often be supplemented with other control measures to achieve satisfactory results.

- Certified seed: Certified seed is checked for the presence of certain seedborne diseases. Therefore, treatments for seedborne pathogens may be unnecessary with certified seed.

- Crop rotation: Crop rotation reduces the populations of many insects and pathogens that survive in soil or crop residue. Seed treatments may be less necessary where crop rotation is practiced.

- Fertility management: Lack of micronutrients, such as chloride, and an excess of major nutrients, such as nitrogen, can favor certain diseases. Maintaining appropriate soil fertility can reduce disease pressure.

- Heat treatment: Hot water treatment can be used to rid seeds of certain seedborne pathogens while leaving the seed viable. For example, the fungi that cause black leg, downy mildew, and anthracnose of cabbage can be eradicated by soaking seed at 122°F for 25 minutes. This treatment will also eliminate the bacteria that cause black rot. Immediately after treatment, seed must be cooled in cold water for several minutes. Then seed must be dried. Procedures must be carefully followed. If the water is too cool, the seedborne pathogens will not be killed. If the water is too warm, the seed may be injured or killed. Because it is difficult and impractical for some seed types, hot water treatment has limited use.

- Planting date: Planting date affects the severity of some root rots, certain insects, and some insect-borne viruses. The classic example is Hessian fly on wheat, which is more likely to occur with early planting. Take-all root rot of wheat, pythium root rot, and barley yellow dwarf are diseases that can be affected by planting date.

- Soil-applied and postemergence sprays: Although seed treatments are convenient and have environmental and economic advantages over soilapplied and postemergence broadcast insecticides and fungicides, for a number of reasons they cannot effectively control every damaging pest.

- Variety resistance: Variety resistance may be available for certain pests. Examples include Hessian fly on wheat, Phytophthora on alfalfa and soybean, and powdery mildew on wheat. Seed treatments may be unnecessary when high varietal resistance is available. For example, use of sweet corn hybrids with high resistance to Stewartís bacterial wilt makes seed treatment insecticides less necessary. However, seed treatments may be an important supplement when resistance is either weak, race specific, or inactive until sometime after seedling emergence.

- Volunteer control: Several insects and diseases use volunteer (self-sown) crop plants as a reservoir. Eliminating volunteer wheat can reduce populations of Hessian fly, aphids, rust, and the like.

Advantages of Seed Treatment

- Seedborne pathogens are vulnerable: The seedborne phase is often the weak link in the life cycle for many plant pathogens. Using seed treatments to control seedborne pathogens is often very effective for disease control.

- Precision targeting: Seed treatments are not subject to spray drift. Because chemicals are applied directly to seeds, little is wasted on nontarget sites, such as bare soil.

- Optimum timing: Seeds and seedlings are generally more vulnerable to diseases and insects than mature plants. Applying treatments to seeds allows pesticides to be present when needed most.

- Low dose: Relatively small amounts of pesticides are used in seed treatments compared to broadcast sprays. This reduces the cost and the potential environmental impact. It also reduces the probability of chemical residues in harvested grain.

- Easy to apply: Seed treatments are relatively easy and cheap to apply compared to broadcast sprays.

Disadvantages of Seed Treatments

- Accidental poisoning: Treated seed looks like food to some animals. Hungry livestock that find carelessly handled treated seed will probably eat it. Birds, such as pheasants or quail, may consume spilled treated seed. Even young children may find and eat improperly stored treated seed.

- Cropping restrictions: Just like other pesticides, some seed treatments may have significant grazing or rotation crop restrictions.

- Limited dose capacity: The amount of pesticide that can be applied is limited by how much will actually stick to the seed. Seed coating technologies are helping to overcome this limitation, but phytotoxicity may still be a problem.

- Limited duration of protection: The duration of protection is often short due to the relatively small amount of chemical applied to the seed, dilution of the chemical as the plant grows, and breakdown of the chemical.

- Limited shelf life of treated seed: Producing excess treated seed is undesirable because the shelf life of treated seed may be limited. Surplus treated seed cannot be sold for grain. This is a particularly serious limitation for seeds such as soybean, where seed germination and vigor decline relatively quickly.

- Phytotoxicity: Pesticide injury to plant tissues is called phytotoxicity. Since seed treatments must exist in high concentrations on the tender tissues of germinating seeds and seedlings, they generally have very low phytotoxicity. A few seed treatments.are partly phytotoxic when applied at high rates. Lower germination and/or stunting may occur if application rates are not carefully controlled. Cracked, sprouted, and scuffed seeds may be particularly susceptible to toxic effects. A few seed treatments may reduce the length of the sprout and, therefore, affect the choice of planting depth.

- Worker exposure: In the course of treating and handling large volumes of seed, workers may be exposed to seed treatment chemicals as aerosols. Inhalation of aerosols and skin contact with seed treatments must be prevented in the seed treatment process.

Seed Treatment Products and Safe use

There are many seed treatment products available, each with different restrictions, labeled uses, active ingredients, dose rates, additives, or formulations. As with most pesticides, each active ingredient has strengths and weaknesses, which is why many seed treatments consist of one or more active ingredients. The degree of pest control often depends on the dose rate of the active ingredient. Some pests may require higher rates than others to achieve control. Some seed treatment labels give a range of rates and indicate pest control responses that are expected for each rate. The applicator must choose the product(s) and rate appropriate for the crop, anticipated pest problem, and the application equipment.

When evaluating a seed treatment, consider the fact that many pathogens and insects are not adequately controlled with current seed treatment products due to one or more of the following product limitations: (1) pesticides with appropriate activity are not available, (2) little or no systemic activity in the plant tissues, (3) limited or no product movement into the expanding root zone, (4) limited product duration, which means peak periods for pest protection and pest infection/damage do not significantly overlap, or (5) effective rates may simply be too expensive or may be phytotoxic to the seed or seedling. Where these limitations exist, pests may be better controlled using genetic resistance, a soil-applied pesticide, or other pest control strategy.

Active Ingredients

Active ingredients are often divided into those that are systemic and those that are non-systemic (contact). Systemic seed treatments penetrate the roots and germinating seed, then move up into stems and leaves. Contact seed treatments protect only the outside of the seed or seedling. Following are some common seed treatment active ingredients, organized by the type of pest protection offered. The information provided here is not complete; be sure to read the product label for activity and current legal uses.

Bactericides

Streptomycin (trade names Ag-Streptomycin and Agri-Mycin) is an antibiotic that kills a broad spectrum of bacteria. It can be used to control seedborne populations of the halo blight pathogen on beans and as a potato seed piece treatment against soft rot and black leg.

Fungicides

Biological agents consist of dormant microorganisms that are applied to seeds. Under favorable conditions, these microorganisms grow and colonize the exterior of the developing seed or seedling. Biocontrol agents may reduce seed decay, seedling diseases, or root rot either by competing with pathogens or by producing antibiotics. Biocontrol organisms include the bacteria Bacillus subtilis (trade name Kodiak) and Streptomyces griseoviridis and the fungus Trichoderma harzianum (trade names T-22, Bio-Trek).

Captan is a broad-spectrum, nonsystemic fungicide effective against various seed decay and damping-off fungi, such as Aspergillus, Fusarium, Penicillium, and Rhizoctonia.

Carboxin (trade name Vitavax) is a systemic fungicide with good activity against smuts and fair activity against general seed rot, damping-off, and seedling blights. It is commonly used to control wheat embryo infections by the loose smut fungus. Carboxin is commonly formulated with other fungicides or insecticides to increase the pest control spectrum.

Difenoconazole (trade name Dividend) is a broadspectrum, systemic fungicide that controls common bunt and loose smut of wheat. At high label rates, it has activity against some fall-season root rots and foliar diseases (powdery mildew and rust). Fall control of root rots and leaf diseases may or may not carry through to the following spring.

Fludioxonil (trade name Maxim) is a broad-spectrum, nonsystemic fungicide effective against various seed decay and damping-off fungi, such as Aspergillus, Fusarium, Penicillium, and Rhizoctonia. In addition, it performs well against seedborne wheat scab.

Imazalil (trade name Flo-Pro IMZ) is a systemic fungicide used against common or dryland root rot of wheat caused by Fusarium and Cochliobolus (also called Helminthosporium). In addition, it performs well against seedborne wheat scab.

Mefenoxam (trade name Apron XL) and metalaxyl (trade names Apron and Allegiance) are closely related, narrow-spectrum, systemic fungicides. They are effective only against Pythium, Phytophthora, and downy mildews. These fungicides are commonly used on a wide range of crops, often in conjunction with a broad-spectrum fungicide, such as captan or fludioxonil.

PCNB (also called pentachloronitrobenzene) is a nonsystemic fungicide. It is especially useful against seedling fungi, such as Rhizoctonia and Fusarium, and has fair activity against common bunt of wheat. PCNB is commonly formulated with other fungicides to increase the disease control spectrum.

Tebuconazole (found in Raxil) is a broad-spectrum, systemic fungicide. It controls common bunt and loose smut of wheat and has activity against some fallseason root rots and some foliar diseases (powdery mildew). Fall control of root rots and leaf diseases may or may not carry through to the following spring. In addition, it performs well against seedborne wheat scab. Tebuconazole is commonly formulated with other fungicides or insecticides to increase the pest control spectrum.

Thiabendazole (also called TBZ) is a broad-spectrum, systemic fungicide useful against common bunt and various seed decay and damping-off fungi, such as Fusarium and Rhizoctonia. In addition, it performs well against seedborne wheat scab. Thiabendazole is commonly formulated with other fungicides to increase the disease control spectrum.

Thiram is a broad-spectrum, nonsystemic fungicide labeled for a wide range of field crops and vegetable crops, and for ornamental bulbs and tubers to control seed, bulb, and tuber decay, and damping-off, as well as common bunt of wheat.

Triadimenol (trade name Baytan) is a broad-spectrum, systemic fungicide that controls common bunt and loose smut of wheat. At high label rates, it has activity against some fall-season root rots and foliar diseases (powdery mildew and rust). Fall control of root rots and leaf diseases may or may not carry through to the following spring. Triadimenol may be formulated with other fungicides to increase the disease control spectrum.

Insecticides

Chlorpyriphos (trade name Lorsban) is a nonsystemic insecticide useful against soilborne insects, such as seedcorn maggot and seedcorn beetle. It belongs to an old class of insecticides called organophosphates, which are currently being phased out by the U.S. Environmental Protection Agency (USEPA).

Diazinon is a nonsystemic, organophosphate insecticide useful against soilborne insects, such as seedcorn maggot and seedcorn beetle. Diazinon has been commonly used as a planter-box treatment but is no longer labeled for soybean.

Imidacloprid (trade names Gaucho and Prescribe) is a systemic insecticide effective against aphids, chinch bug, flea beetle, Hessian fly, leafhopper, seedcorn maggot, thrips, whitefly, white grubs, and wireworms. It reduces incidence of some diseases by controlling the insect vectors. Length of control is influenced by the dosage used. For example, high label rates may be needed to reduce potential spread of barley yellow dwarf virus due to aphid vectors. Research has shown that imidacloprid seed treatment provides limited control of corn rootworms. Where the potential for high corn rootworm pressure exists, a soil-applied insecticide should be considered.

Lindane is a nonsystemic insecticide useful against soilborne insects, such as wireworm and seedcorn maggots. It belongs to an old class of insecticides called chlorinated hydrocarbons, which are currently being phased out by the USEPA. Lindane has been commonly used as a planter-box treatment but has largely been replaced by newer pyrethroid insecticides.

Permethrin (trade names Barracuda, Kernel Guard Supreme, and Profound) is a nonsystemic, pyrethroid insecticide useful against soilborne insects, such as wireworm and seedcorn maggots.

Tefluthrin is the active ingredient found in Force soil insecticide. As a seed treatment (marketed as Proshield), tefluthrin is currently only available on certain field corn hybrids. It is a nonsystemic, pyrethroid insecticide useful against soilborne insects such as wireworm and seedcorn maggots. Research has shown that Proshield seed treatment provides limited control of corn rootworms. Where the potential for high corn rootworm pressure exists, a soil-applied insecticide should be considered.

Thiamethoxam (trade name Cruiser) is a systemic, neonicotinoid insecticide effective against various sucking and chewing pests, such as thrips, aphids, Colorado potato beetles, seedcorn maggot, Hessian fly, flea beetles, leafhoppers, chinch bugs, and wireworms. Currently, the product is labeled for use only on wheat and barley seed.

Formulations

Seed treatment pesticides are commonly formulated as a dry flowable (DF), flowable (F), flowable seed treatment (FS), liquid (L), liquid suspension (LS), or wettable powder (WP). Although the formulation you use may seem trivial, it can have a major impact on your equipment and treatment uniformity. For example, some formulations may not mix well in the tank, and some may readily settle out without constant agitation. Some products come in water-soluble packaging, which benefits the applicator by reducing exposure to the pesticide. However, as with all formulations, be sure to read and follow the label instructions to ensure proper mixing and compatibility.

Additives

Seed treatment products usually contain a variety of additives in addition to the active ingredients. If important additives are lacking in a product, they often can be added to the pretreatment mixing tank. Before using additives, consult the manufacturerís instructions to avoid problems and duplication.

Colorants or dyes are added to mark treated seed and prevent mixing with food grain. Colorants improve the appearance and also help ensure uniformity of treatment coverage. Color-enhancing agents may be added to further improve the appearance.

Carriers, binders, and stickers are listed on the label as inert ingredients. There is no requirement that the name of these materials be given. They are selected by the manufacturer; approved by the USEPA; and are usually neutral in pH, nontoxic to humans, and cause no apparent damage to the germination of the seed. They are added to increase the adherence of the pesticide to the seed, prevent dusting off, and/or cut down the dustiness in the seed treatment facility.

Antifoam agents suppress formation of troublesome foam. Lubricants, such as graphite or talc, reduce the friction of seed flow through the planter. Micronutrients, such as molybdenum, may be added to soybean seed treatments as a convenient way to introduce trace elements required for nodulation.

Compatibility Issues

When using unfamiliar mixtures or rates (or even when using familiar mixtures and rates with an unfamiliar crop), be sure to read and follow the labels of each product, To test the mixture for physical compatibility, make a small slurry (include all products in the correct ratio) and observe for signs of incompatibility, such as settling, separation, gelling, or curdling. Furthermore, check the germination of a small amount of seed before committing the total seed lot to a selected treatment.

Do not assume that biological seed treatments will be compatible with chemical seed treatments; contact the manufacturer of the biological seed treatment product for questions about compatibility. Finally, if a nitrogenfixing inoculant will be used In most cases, the inoculant should be applied just before plant-ing, as the beneficial bacteria may not survive extended contact with certain pesticides. In-furrow application of inoculants may allow for use of seed treatments otherwise considered incompatible or marginally compatible.

Labeling and Visually Identifying Treated Seed

The Federal Seed Act and the Oregon Seed Law require special labeling for treated seed. The labeling provides the end user with instructions on proper handling and storage of treated seed. It is a practical means of ensuring that treated seed is not used for any purpose other than for planting. In addition, the Occupational Safety and Health Administration (OSHA) may also require that seed treaters provide additional safety and training to help protect the health of persons working in and around seed treatment facilities or those handling treated seed.

Labeling Treated Seed

Federal and state seed laws mandate that seed treaters print the following specific information about the treatment on the same tag bearing the analysis information, on a separate tag attached to the seed container, or directly on a side or the top of the container:

- A word or statement indicating that the seed is treated, such as "Treated".

- The accepted common or chemical name of all pesticides applied.

- A statement, such as "Do Not Use for Food, Feed, or Oil Purposes".

- In rare cases where a highly toxic pesticide (indicated by the "Danger/Poison" signal word on the pesticide product label) is applied to seed, the seed tag must also bear a skull and crossbones and a precautionary statement, such as "Treated with Poison." The word "poison" must be in type no smaller than 8 points and shall be in red letters on a distinctly contrasting background. In addition, the skull and crossbones must be at least twice the size of the type used for the precautionary statement.

This Seed Treated With Captan

Do not use for food, feed or oil purposes

Basic labeling for treated seed.

OSHA has implemented additional label requirements under the hazard communication standards. This information is usually provided with the labeling for the treatment product and may include:

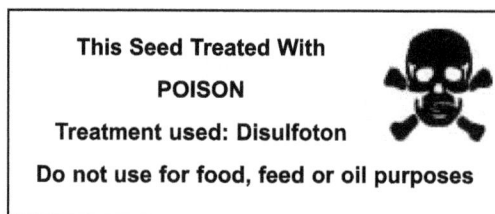

This Seed Treated With

POISON

Treatment used: Disulfoton

Do not use for food, feed or oil purposes

Basic labeling for seed treated with a highly toxic pesticide

- Appropriate hazard warnings (for example, statements such as "This chemical is a skin irritant" or "This chemical is capable of causing irreversible eye damage or other adverse effects").

- The name, address, and phone number of a responsible party to contact in case of problems.

The information described is a minimum requirement. Some seed treatment pesticide labels recommend additional information that the treater should add to the label of treated seed. This may include special personal protection information, environmental hazards, statements of practical treatment in case of an accident, or grazing restrictions.

Remember, the only legal use for treated seed is planting. Seed treatment pesticide labels prohibit the use of treated seed for food, feed, or oil purposes.

Coloring Treated Seed

Federal Food and Drug Administration regulations require that all food grain seeds (for example, wheat, corn, oats, rye, barley) treated with seed treatment pesticide formulations be colored with an approved dye to contrast with the natural color of the seed to prevent its use as food or feed. This requirement provides a convenient means of detecting the presence of treated seed mixed with food or feed grains or products. Coloring is not required for planter-box formulations; however, dyes are commonly used to help the user confirm proper coverage.

Most seed treatment pesticides for in-plant use come from the manufacturer with the dye added. However, some seed processors may apply additional dye to modify the color. Approved colorants include dyes, color coat pigments, or color films. Dyes are designed to stain the seed with color, color coat pigments cover the seed with color, and color films are made from a polymer that creates a colored film around the seed. There is a wide variety of color and surface texture options available to seed processors. In addition to bright contrast, colorants must not affect seed germination or pose a health threat to personnel processing or using the seed.

Shipments of food grain seeds treated with a pesticide formulation lacking a colorant or grains that contain a mixture of treated and nontreated seed are subject to seizure and possible destruction. In addition to financial loss to the owner of the shipment, the person responsible for the violation may be prosecuted.

Safe Handling Practices

Handle seed treatment pesticide products with care. Product labels provide information on safe handling and application. Always read the label and follow instructions precisely. The label also provides the applicator with information about first aid, potential environmental hazards, directions for use, proper storage, and container disposal.

Managing a Safe Seed Treatment Facility

Isolate the seed treatment area from other facility functions to keep pesticide dust and fumes from reaching unprotected employees and stored agricultural commodities. Install an approved exhaust and dust collecting system to remove toxic vapors and dust from the operating area.

For legal and safety reasons, it is important that nontreated seed does not become contaminated with pesticide residue and colorant. Sacks, containers, trucks, wagons, augers, and conveyors used for transport of treated seed should be used for that purpose only. Be sure to properly dispose of contaminated sacks so that they are not used for any other purpose.

Thoroughly clean seed treatment equipment between different product batches to avoid cross contamination. Consult the pesticide manufacturer for the names of appropriate cleaning techniques and directions. Regularly cleaning seed treatment equipment also can help prevent equipment corrosion and settling or clogging problems.

For minor spills or leaks, follow all instructions indicated on the label and material safety data sheet (MSDS). Clean up spills immediately. Take special care to avoid contamination of equipment and facilities during cleanup procedures and disposal of wastes. For major spills, Oregon law requires that emergency notification be made to the Oregon Emergency Responce System (OERS). They in turn notify the appropriate agencies for response.

Personal Safety

Use caution when handling seed treatment chemicals. Remember that exposure to seed treatment pesticides may cause a wide range of acute and chronic toxic reactions in people. When handling seed treatment pesticides:

- Read and become familiar with the label and MSDS for each pesticide that you use. Make certain that these documents are readily available at all times, and refer to them in the event of an accident. Labels do change (often with no notice), so be sure to review the label each time you purchase a pesticide.

- Avoid inhaling pesticide dust or vapor, and always protect skin and eyes from exposure. Use proper protective equipment recommended by the pesticide label. Consider wearing goggles, rubber gloves, and a rubber apron, even when the product label does not specifically require it.

- Wash thoroughly with soap and water before eating or smoking.

- In case of exposure, immediately remove any contaminated clothing and wash the affected area thoroughly with soap and water. For safety purposes, install a safety shower in the immediate vicinity of the treatment equipment.

- When treating large amounts of seed, change clothing frequently enough to avoid buildup of pesticides in the garments.

- No matter how tired you may be, shower immediately after work and change all clothing. Wash clothing thoroughly (separate from the family wash) before reuse.

Proper Storage and Disposal

Pesticides

Store unused seed treatment pesticides in their original, labeled, tightly closed containers in a dry, ventilated, locked location inaccessible to animals, children, and untrained persons. Do not store pesticide products near heat, fire, sparks, or open flame, or in direct sunlight. Protect liquid formulations from freezing temperatures. Check the pesticide label for appropriate product storage directions. Do not reuse pesticide containers.

Completely rinse and puncture empty pesticide containers, and offer for recycling or dispose of them in a sanitary landfill. Consider recycling excess pesticide solution and treatment wastewater by using them to dilute the next batch of the same product. Do not divert wastewater into ponds, lakes, streams, ditches, sewers, drains, and the like. Handle wastewater as you would excess pesticide. Always follow all local, state, and federal pesticide disposal requirements.

Treated Seed

Pesticide-treated seed must be stored in a dry, wellventilated location separate from untreated seed; it should never be stored in bulk storage bins that might also be used for edible grain storage. Store treated seed in special multiwall (3- or 4-ply) or tightly woven bags. Some polyethylene or foil-lined bags are also good containers for treated seed. Make sure seed is thoroughly dry before bagging, as excessive moisture can cause rapid deterioration of the seed. Clearly label the seed to indicate the type of seed treatment. If it is held in storage for a year or more, check the germination percentage prior to sale.

Very few pesticide labels provide useful guidance regarding disposal of treated seed. As a result, proper and legal disposal of unwanted treated seed has been a contentious issue for many years. Certainly seed treatment pesticide labels prohibit the use of treated seed for food, feed, or oil purposes. Thus, it is illegal to mix treated seed with food or feed products in an attempt to get rid of excess or otherwise unwanted treated seed. Obviously treated seed may be disposed of by planting it at an agronomically acceptable seeding rate. However, surface application without incorporation may present a hazard to humans and animals and may be illegal.

SEED ENHANCEMENT

Seed enhancements include physical, physiological and biological treatments to overcome germination constraints by uniform stands, earlier crop development and better yields. Improved germination rates and seedling vigour are due to reduced emergence time by earlier start of metabolic activities of hydrolytic enzymes and resource mobilization. Nutrient homeostasis, ion uptake, hormonal regulation, activation of antioxidant defence system, reduced lipid peroxidation and accumulation of compatible solutes are some mechanisms conferring biotic and abiotic stress tolerance. Several transcription factors for aquaporins, imbibitions, osmotic adjustment, antioxidant defence and phenylpropanoid pathway have been identified. However, the knowledge of molecular pathways elucidating mode of action of these effects, reduced longevity of primed or other physical and biological agents for seed treatments and market availability of high-quality seeds are some of the challenges for scientists and seed industry. there is need to minimize the factors associated with reduced vigour during seed production, improve seed storage and handling, develop high-tech seeds by seed industry at appropriate rates and integrate agronomic, physiological and molecular seed research for the effective regulation of high-quality seed delivery over next generations.

Seed Enhancements

Seed enhancements or seed invigoration are the post-harvest treatments used for improving the germination and growth of seedlings required at the time of sowing. Many shotgun approaches are being used for seed enhancement for the last 24 years, which includes seed priming, magnetic stimulation, seed pelleting and coating.

Seed Enhancement using Physical Agents

Physical treatments are applied externally without any hydration or application of chemical materials to the seeds. The main purpose is to enhance germination and seedling establishment. The mechanism of seed invigoration with physical techniques is still unknown. The work on exposure of seeds to radiation was started in early 1980s and now a number of studies have been focused on the use of plasma technology for seed invigoration of agronomic and horticultural crops. Magnetic field treatments are being considered as effective seed enhancement tools for agronomic and horticultural crops; however, their application is limited at large scale. Among physical methods, magnetic field and irradiation with microwaves or ionizing radiations are the most promising pre-sowing seed treatments. Thus, physical seed enhancements are an alternative approach to other chemical seed invigoration treatments, which provide better solution for the growing world seed market.

Magnetic Fields for Seed Treatments

Magnetic seed stimulation involves identifying the magnetic exposure dose to affect the germination, early seedling growth and subsequent yield of crop plants. The magnetic exposure dose is the product of the flux density of magnetic field and of the time to exposure. The flux density of magnetic field varies with static or alternating magnetic fields exposure to seeds. Magnetic field ensures the quick germination, uniform crop stand establishment and yield of many agronomic and horticultural crops. These not only increase the rate of germination, growth and yield but also reduce the attack of pathogenic diseases.

Magnetic field exposure increases the germination of non-standard seeds and also improves their quality. Magnetic field influences the initial growth stage of the plants after the germination. In recent years, work on magnetic-treated water revealed that plant growth and seed germination were improved by priming.

Plasma Seed Treatments

Plasma application in agriculture and medicine is a recent advancement. The agricultural aspects include seed germination and plant growth. Many researches report that germination and growth enhancement mechanism is affected by use of plasmas with several gases as aniline, cyclohexane and helium . To enhance seed development and plant-growth microwave plasma, magnetized plasma and atmospheric plasma are adopted treatments. The effect of gases is much commonly studied in plasmas treatments. Various reports revealed that the quality of plant development controlling thiol groups is diversified by redox reaction persuaded by the active oxygen species of water vapour plasma.

Non-thermal plasma radiations are applied in agriculture as alternative to scarification, stratification and priming helped to improve the plant growth. Plasma helps to attain zero seed destruction, no chemical use and environment friendly treatments to seeds. Plasma treatment improves seed quality and plant growth. Seed exposure to plasma also resulted in alterations of enzymatic activity and caused sterilization of seed surface.

Plasma chemistry can tune seed germination by delaying or boosting with application of plasma-treated deposits on seed surfaces. The recent important plasma-related investigation includes the practice of microwave discharges and low-density radio frequency (RF) discharges. The discharge of atmospheric pressure and the discharge of coplanar barrier have been assessed in recent studies. The investigation of various seed germination patterns was implemented on different seeds including wheat, maize, radish, oat, safflower and blue lupine. Safflower seeds expressed 50% greater germination rate when treated with radio frequency plasma for 130 min with argon. Soybean seeds were treated with cold plasma treatment with 0, 60, 80, 100 and 120 W for 15 s and found positive effects of cold plasma treatments on seed germination and seedling growth of soybean.

Radiation Seed Treatments

With recent advancements in agriculture, gamma radiations can improve plant characteristics such as precocity, salinity tolerance, grain yield and product quality in suboptimal environment depending upon the level of irradiation. Second, gamma radiation can also sterilize agricultural products to prevent pathogen infestation thus increasing conservation time during storage and trading.

The biological effects of radiations is based on chemical interaction with biomolecules and water to produce free radicals that can manipulate biomolecules and induce cell to switch on antioxidant system that prepared the defensive shield against upcoming stresses. In spite of the conventional seed enhancements, physics has manipulated radiation dose to trigger biochemical reactions necessary for seed germination without affecting seed structural integrity and collateral DNA damage. It was found that the low dose of gamma radiation (up to 20 Gy) on germination of three varieties of Chinese cabbage shows a positive impact.

Physiological Seed Enhancements

Seed Priming

Seed priming is a pre-sowing approach for influencing the seedling development by stimulating pre-germination metabolic activities prior to the emergence of radicle and improvement in the germination rate and performance of plant. Seed priming is a controlled hydration process in which seeds are dipped in water or any solution for a specific time period to allow the seed to complete its metabolic activities before sowing and then re-dried to original weight.

Priming treatments include osmopriming by polyethylene glycol (PEG) or a salt solution, hydropriming solid matrix priming in which seeds are soaked in inert medium of known matrix potential and hormonal priming. A balance of water potential between osmotic medium and seed is necessary for conditioning, and different non-penetrating agents such as organic solutes and salts are used for this purpose. Furthermore, these priming treatments show positive response only at sub-optimal or supra-optimal field conditions such as drought, excessively high or low temperatures and salinity.

Hydropriming

Hydropriming is a controlled hydration process that involves seed soaking in simple water and then re-drying to their initial moisture. No chemical is used during this technique but some cases of non-uniform hydration causes uneven germination . Among the different seed enhancement techniques, hydropriming could be a suitable treatment under salinity stress and drought-prone environments .

Hydropriming as a risk free, simple and cheap technique has become popular among farmers, with promising effects in the context of extensive farming system. Hydro-primed seeds produced healthy seedlings, which resulted in uniform crop stand, drought resistance, early maturity and somewhat improved yield.

Osmopriming

Osmopriming involves seed hydration in an osmotic solution of low water potential such as polyethylene glycol or a salt solution under controlled aerated conditions to permit imbibition but prevent radical protrusion. For osmopriming, mostly polyethylene glycol or salt solution is used to regulate water uptake and to check radicle protrusion. Most commonly used salts for osmopriming are potassium chloride (KCl), potassium nitrate (KNO_3), sodium chloride (NaCl), magnesium sulphate ($MgSO_4$), potassium phosphate (K_3PO_4), calcium chloride ($CaCl_2$) and potassium hydrophosphate (KH_2PO_4). All these salts provide nutrient like nitrogen to the germinating seed, which is required for the protein synthesis during the germination process. However, these salts rarely cause nutrient toxicity to the germinating young seedlings. Osmopriming induced more rapid and uniform germination and resulted in decreased mean germination time.

Hormonal Priming

Plant-growth hormones or their derivatives contained by several products are indole-3-butyric acid (IBA), an auxin and kinetin type of cytokinin. Cytokinins play a vital role in all phases of plant development starting from seed germination up to senescence . Priming with optimum concentration of cytokinins has been reported to increase germination, growth and yield of many crop species. Gibberellic acid (GA_3) is known to break seed dormancy, enhance germination, hypocotyl growth, internodal length, and cell division in the cambial zone and increase the size of leaves. GA has stimulatory effect on hydrolytic enzymes, which speed up the germination and promote seedling elongation by degrading the cells surrounding the radicle in cereal seeds.

Various naturally occurring growth promoting substances such as moringa leaf extract, chitosan, sorghum water extract and seed weed extract are commonly used for seed priming. Moringa (*Moringa oleifera* L.) as a natural source of plant-growth regulators contains cytokinins as zeatin . In addition, moringa leaf extracts contain higher concentrations of various growth enhancers such as ascorbates, phenolic compounds, K, and Ca. Priming maize seed with moringa leaf extract reduces mean germination (MGT) and T_{50} with increased germination index and germination count that ultimately improved seedling growth by increasing chlorophyll content, amylase activity and total sugar contents under chilling conditions. Moringa leaf extract diluted up to 1:36 with water was applied on various field crops and 35% increase in the yield of sugarcane, sorghum, maize, turnip and bell pepper was observed. Nonetheless, moringa leaf extracts being low cost can be a viable option for improving the productivity of resource poor farmers.

Nutrient Priming

The application of micronutrients with priming can improve stand establishment, growth and yield; furthermore, the enrichment of grain with micronutrients is also reported in most cases. Many researchers proved the potential of nutrient priming in improving wheat, rice and forage legumes. Among micronutrients, Zn, B, Mo, Mn, Cu and Co are highly used as seed treatments for most of the field crops.

Seed treatment with micronutrient is a potentially low-cost way to improve nutrition of crops. Farmers have responded in South Asia in a positive way in the seed treatment, which is a simple technique soaking seeds in water overnight before planting. Seed priming with zinc salts is used to increase growth and disease resistance of seedlings.

Biological Seed Enhancements

Bacterial Seed Agents

Plant-growth-promoting rhizobacteria (PGPR) are free-living, soil-borne bacteria, which when applied to soil, seeds or roots promote the growth of the plant or reduce the incidence of diseases from soil-borne plant pathogens. PGPR can influence plant growth either directly or indirectly through fixation of atmospheric nitrogen, solubilization of phosphorus and zinc and producing siderophores, which can solubilize/sequester iron, synthesize phytohormones, including auxins, cytokinins and gibberellins to stimulate plant growth, and synthesize ACC-deaminase enzyme by modulation of ethylene level under stress conditions.

Among various genera of PGPR endophytes are good priming agents because they colonize roots and create a favourable environment to develop and function with their hosts—symbiotic partner.

Biopriming is a new technique of seed enhancement integrating biological (inoculation of seed with beneficial organism to protect seed) and physiological aspects (seed hydration) to promote plant growth, development and suppression of diseases. It is used as an alternative approach for controlling many seed- and soil-borne pathogens. Seed priming with beneficial microorganisms (bacteria and fungus) often result in more rapid growth and increase plant vigour and may be useful under adverse soil conditions. Besides diseases control, the application of PGPR as a biopriming agent for biofertilization is an attractive option to reduce the use of chemical fertilizers. PGPR that have been tested as co-inoculants with rhizobia include strains of the following rhizobacteria: *Azotobacter*, *Azospirillum*, *Bacillus*, *Pseudomonas*, *Serratia* and *Streptomyces*.

Role of a Bacterial Biopriming Agent in Plant-Growth Promotion

The Plant growth promoting bacteria (PGPB) are a heterogeneous group of beneficial microorganisms present in the rhizosphere, on the root surface or inside plant tissues, and are able to accelerate the growth of plants and protect them from biotic and abiotic stresses. Beneficial effects of biopriming have been reported in several vegetable seeds. Priming of tomato seed with beneficial bacteria improved the rate of germination, seedling emergence and growth of plant. The beneficial response of biopriming on seed germination and seedling vigour in chilli was reported . Similarly, improvement in okra growth and yield was reported up to 60% when seeds were bioprimed with *P. fluorescens* culture. In experiments where lettuce plants were treated with *Bacillus* strains, it was observed that after two weeks the tissues of roots and shoots contained a greater amount of cytokinin than control plants. The accumulation of cytokinins was associated with a 30% increase in plant biomass

Role of a Bacterial Biopriming Agent in Plant Disease Control

Seed enhancement by biopriming agents involves coating/soaking the seed with one biological agent or microbial consortium, then incubating the seed under optimum (temperature, moisture) conditions.

Biopriming of seeds with different bacterial strains particularly rhizobacteria have been shown to be effective in suppressing disease infection by inducing a resistance mechanism called 'induced systemic resistance' (ISR) in varied agronomic and horticultural crops. Among various bacterial genera, *Bacillus* and *Pseudomonas* spp. are ubiquitous rhizosphere inhabitant bacteria that are the most studied biopriming agents reported as disease suppressing in plants. Priming seeds of many crops with biological control agents (BCA), *Bacillus subtillus* and *Pseudomonas fluorescens* are the most effective approach for controlling seed and root rot pathogens and as a substitute for chemical fungicides without any risk to human, animal and the environment.

Seed Enhancement by Alleviating Abiotic Stresses using Biopriming

Seed priming with beneficial microorganisms may promote plant growth and increases abiotic stress tolerance in arid or semiarid areas. PGPB are adapted to adverse conditions and protect plants from the deleterious effects of these environmental stresses, thus increasing crop productivity. Bioprimed seeds with *Enterobacter* sp. P-39 showed maximum improvement in germination and seedling growth of tomato under osmotic stress. Shows the selected examples of beneficial response of biological inoculants for enhancing growth and yield of various crops under normal and stress conditions.

Fungal Seed Agents for Biopriming

In this approach, beneficial bacterial and fungal agents are exploited for the purpose of biopriming of seeds to enhance growth, yield and mitigation of biotic and abiotic stresses. It is an environmental friendly, socially accepted approach and also offers an alternative to the chemical treatment methods gaining importance in seed, plant and soil health systems. Seed biopriming enhanced drought tolerance of wheat as drought-induced changes like photosynthetic parameters and redox states were significantly improved by *Trichoderma* sp. under stress conditions over control. Very recently, Junges et al. compared the potential of biopriming (*Trichoderma* and *Bacillus* spp.) with commercial available products Agrotrich plus and Rhizoliptus for enhancing growth and yield of beans. Results revealed that biopriming with spore or bacterial cell suspensions promoted bean seedling growth compared to other techniques.

Seed Coating and Pelleting

Seed film coating, pelleting, priming and inoculation are globally practiced seed treatments used with the objectives of enhancing plantability, distribution, germination and storage of seeds. These techniques aim to apply adhesive films, fungicides, herbicides, growth promoters and biological agents. Seed coating is carrier of chemical materials to support seedling growth. Compounds such as growth regulators, inoculants, micronutrients, fungicides, insecticides and other seed protectants are applied to the pellet to enhance seed performance.

Seed coating demands uniform application of inert material over the seed surface. This also helps to protect the seed from soil and seed-borne pathogens. Pharmaceutical industry uses seed polymer coating for a constant application of numerous materials to seeds. The commercially available plasticizers, polymers and colourants (commercially they are readily available to be used as liquid) are applied as film formulations. However, the exact composition of coating material is a carefully guarded secret by the companies who develop them. Usually, coating material contains binders, fillers (e.g., polyvinyl alcohol, gypsum and clay) and an intermediate layer (e.g., clay, polyvinyl acetate and vermiculite). Seed agglomeration is an alternate coating technology with the purpose to sow multiple seeds of the same seed lot, or multiple seeds of different seed lots, varieties or species.

Mechanisms of Seed Enhancements

Physiological and Biochemical Aspects

Improved crop performance through pre-sowing treatments depends on the nature of compounds used for priming and their accumulation under abiotic stresses. These compounds include inorganic salts, osmolytes, phytohormones, tertiary amino compounds such as glycinebetaine, amino acids and sugar alcohols including bioactive compounds from micro-organisms. For instance, the application of compatible solutes as seed priming improves salinity resistance by cytosolic osmotic adjustment indirectly by enhancing regulatory functions of osmoprotectants. Chilling-induced cross-adaptation salt tolerance in wheat is associated with enhanced accumulation of beneficial mineral elements (K^+ and Ca^{2+}) in the roots and reduced uptake of toxic Na^+ in the shoots through ionic homeostasis and hormonal balance with greater concentrations of indoleacetic acid, abscisic acid, salicylic acid and spermine in chilled wheat seeds. In flooded soils, improved stand establishment in rice through seed priming is related to enhanced capacity of superoxide dismutase (SOD) and catalase (CAT) activities to detoxify the reactive oxygen species in seeds and greater carbohydrate mobilization. These effects are more pronounced in tolerant genotypes that emphasize to combine crop genetic tolerance with appropriate seed treatments to improve seedling establishment of rice sown in flooded soils.

Such enhanced remobilization efficiency in seed embryos of cereals coated with hydro-absrobers is related to change in activities of enzymes for sucrose breakdown upon moisture absorption. Coated seeds absorb more moisture that creates anoxic conditions in developing embryos but genetic difference are found for sucrose breakdown in rye, barley and wheat with change in invertase activities due to difference in timing of imbibitions.

Beneficial effects of magnetic seed stimulation are associated with various biochemical, cellular and molecular events. Pre-sowing magnetic seed treatment also increases ascorbic acid contents by stimulating the activity of the enzymes and proteins. Physiological and biochemical properties also increase due to enhanced metabolic pathway by the free movement of ions. However, its biochemical and physiological mechanisms are still poorly understood.

Molecular Aspects

Favourable effects of priming at cellular level include RNA and protein synthesis. Seed priming induces several biochemical changes within the seed needed for breaking seed dormancy, water imbibition, enzymes activation, hydrolysis of food reserves and mobilization of inhibitors. At cellular level priming initiates cell division transportation of storage protein. Higher germination rate and uniform emergence of primed seed is due to metabolic repair with increased production of metabolite required for the germination during the imbibition process. Priming increased the production and activity of α-amylase within germinating seeds, thus increased the seed vigour.

Several proteins and their precursors for regulation involved in different steps of seed germination or priming have been identified using model plant Arabidopsis. The expression of these proteins such as actin isoform or a WD-40 repeat protein occurs in imbibition and cytosolic glyceraldehyde-3-phosphate dehydrogenase in the seed dehydration process. Priming-induced changes in proteins levels have been identified as peroxiredoxin-5, 1-Cys peroxiredoxin, embryonic protein DC-8, cupin, globulin-1 and late embryogenesis abundant protein. The expression of these proteins led to improved seed germination and the expression of these embryo proteins remained unchanged even after priming.

A major quantitative trait locus (QTL) Htg6.1 of seed germination responsive to priming under high temperature stress using a recombinant inbred line (RIL) of lettuce has been identified. The expression of this QTL at high temperature is coded by a gene *LsNCED4* encoding the key enzyme, i.e., 9-cis-epoxycarotenoid dioxygenase, of the abscisic acid biosynthetic pathway and maps precisely with Htg6.1. However, *LsNCED4* gene expression was higher in non-primed seeds after 24 h of imbibition at high temperature compared to the expression of *LsGA3ox1* and *LsACS1* genes encoding enzymes of gibberellins and ethylene biosynthetic pathways, respectively. *LsNCED4* gene expression was reduced after priming and when imbibition was carried out at the same temperature . The seed response to priming in terms of germination and temperature sensitivity is associated with temperature regulation of hormonal biosynthetic pathways.

Osmopriming induced quantitative expression of stress-responsive genes such as CaWRKY30, PROX1, Osmotin for osmotic adjustment, Cu/Zn SOD for antioxidant defence and CAH for phenylpropanoid pathway. The same genes were induced earlier or at higher levels in response to thiourea priming at low temperature. The expression of these genes imparts cold tolerance in capsicum seedlings. Notably, high levels of other plant-growth hormones, such as indolyl-3-acetic acid (IAA) and abscisic acid (ABA), were also observed. The authors suggested that *Bacillus* strains have dual effect on plant-growth promotion and accumulation of cytokinins by increasing other routes of synthesis of hormones such as IAA and ABA, as well as interfering in other hormonal balance synthesis such as gibberellins (GA). Using advanced molecular tools such as proteomics may help to detect protein markers that can be used to unravel complex development process of seed vigour of commercial seed lots, or analysis of protein changes occur in industrial seed priming treatments to accelerate seed germination and improve seedling uniformity.

Seed Enhancements and Plant Development

Modulation of Seedling Growth

Seedling vigour is important to help ensuring good crop establishment. Pre-sowing seed treatments offer pragmatic solution to poor seedling establishment by overcoming the germination constraints under normal and adverse conditions. Several researches have shown the potential of chemical priming, use of macro- and micronutrients, natural compounds of plant origin and plant-growth-promoting bacteria including water under greenhouse and field conditions. Most of the priming techniques such as osmopriming and on-farm priming have been optimized for specific crops for soaking duration and concentration. For chemical priming, polyamines including spermine, spermeidine and putrescine, calcium chloride ($CaCl_2$), potassium chloride (KCl), NaCl, KH_2PO_4, KNO_3, PEG, hydro-absorbers such as humic acid and biplantol for seed coating and naturally occurring molecules such as nitric oxide (NO), hydrogen sulphide (H_2S), H_2O_2, ascorbate, salicylic acid, indoleamine molecule melatonin (Mel) and most recently growth promoting cytokinin-rich moringa leaf extracts are commonly being evaluated. The endogenous levels of naturally occurring molecules when applied as seed priming may increase initially and later with subsequent improved growth.

The beneficial effects of seed priming have been documented in cereals, sugar crops, oilseeds and horticultural crops. Early seedling growth by pre-sowing seed treatments is due to improved germination rate, reduced time of germination or emergence, and uniform and enhanced germination percentage contributed by enhanced mobilization of germination metabolites from endosperm towards growing embryonic axis. However, variation in germination rates with seed coating thickness and composition has been found which ultimately affects the mobilization efficiency of seed reserves. Therefore, the use of hydro-absorbers is suggested for coated seeds to enhance the efficiency of germination metabolites which may differ among species.

Seed priming with nutrients usually increases the seed contents of primed nutrients, which may be translocated to the growing seedling to support the seedling development. Improved seedling growth and dry mass may be attributed to enhanced nutrient uptake and enzymes associated under deficient conditions and offer perspective for improved seed quality at crop harvesting. Priming mediated by manganese (Mn) has also significant effect on the growth and yield performance of crops. In comparison to soil application, Mn priming improved stand establishment, growth, yield and grain contents.

The concentration of these nutrients may be toxic when used in relatively higher concentration. For instance, priming with 0.5% Boron solution completely suppressed the germination and growth in rice and 0.1 M $ZnCl_2$ and 0.5 M $ZnSO_4$ in wheat. Seed priming induced early vigour indices have been associated with suppression of weeds in primed stand of aerobic rice. Germination, shoot biomass and total root length were increased in seeds of cultivar IR74 containing Pup1 QTL after water priming. This suggests that seed management approaches may be combined with genetics to improve the crop establishment in different crops including rice under P-deficient conditions.

Pre-sowing magnetic seed treatment of wheat seeds has an effect on the germination, and the growth rate was increased to 23% while the germination rate was 100% in the laboratory and less time was taken with 15 min treatment.

Effects on Crop Phenology

Plants grown by primed seeds usually emerge faster and complete other developmental stages such as tillering, flowering and physiological maturity earlier than seeds without priming. This developmental plasticity of priming may be beneficial when crop planting is delayed due to adverse climatic conditions such as low temperature or high rainfall at sowing, high temperature at reproductive stage and may help plant to avoiding detrimental conditions by earlier maturity without yield decrease. In fact, earlier and vigorous crop stand usually captures more resources of water and nutrients through better root system and had larger leaf area and duration with enhanced photo-assimilation that subsequently contributes towards better yield. However, integrated studies combining seed priming with other crop husbandry practices such as planting geometry, irrigation and fertilization may be interesting in crop stress and nutrient management for improved resource use efficiency.

Yield Improvement

Seed priming benefits are not usually end up with improved crop stand. Several studies report long-lasting effects on yield-associated advantages in terms of increased growth rates, high dry matter production and produce quality by improving crop resistance to biotic and abiotic stresses. A very few reports showing no yield improvement by seed priming are available. Seed priming improved yield is due to reduced weed biomass, higher leaf area index and panicles/m^2 in aerobic and submerged rice, respectively, improved crop nutritional status of nutrients primed in maize under low temperature stress, comparatively better dry matter production with higher tissue Zn concentration with Zn seed priming in rice, reduced spikelet sterility in direct seeded rice irrigated with alternate wetting and drying (AWD) and under system of rice intensification (SRI) condition with improved crop growth and higher tillering emergence. Likewise, early planting spring maize stimulated seedling growth due to increased leaf area index, crop growth and net assimilation rates, and maintenance of green leaf area at maturity, better stand establishment in no tilled wheat under rice-wheat system, with enhanced tillering emergence and panicle fertility and with B nutrition under water saving rice cultivation, GA_3 priming induced modulation of ions uptake (Na^+, K^+) and hormonal homeostasis under salinity in wheat, in combination with gypsum + FYM treatment by ameliorating effects on plant growth and improving performance of poor quality wheat seeds under drought stress. Nonetheless, another researcher observed improved yield due to stand establishment and increasing panicle number by coating rice seeds with Zn-EDTA or ZnO or Zn lignosulfonate.

Crop emergence, crop growth and development of two pea varieties with a significant increase in the seed yield have been reported. It was reported that contents of sugar were increased with magnetic seed stimulation in sugar beet roots, and gluten contents were also increased in wheat kernals when magnetic field was applied to the seeds before sowing. Similarly, many researchers had reported higher grain yield due to improved stand establishment, growth and development in agronomic and horticultural crops.

Nonetheless, priming effects are not only limited to stand establishment and yield, water productivity and uptake of beneficial minerals with reduction in harmful ion but the quality of harvested produce is also improved. Thus, it offers promising and economical solution to improve crop resistance against low and high temperature, flooding and drought, salinity and nutrient stress and effective strategies for agronomic biofortification when combined with soil management and crop genetics.

References

- Seeds, classes-of-seeds, agriculture, agri-inputs: vikaspedia.in, Retrieved 25 May, 2019

- Seed, types-of-seed, types-of-seed-dicotyledonous-and-monocotyledonous-seeds-48811: biologydiscussion.com, Retrieved 10 April, 2019

- Plants, seeds-and-their-morphological-features-with-diagram-6334: biologydiscussion.com, Retrieved 06 June, 2019

- Seedling, definition-214: maximumyield.com, Retrieved 27 July, 2019

- Recent-advances-in-seed-enhancements, new-challenges-in-seed-biology-basic-and-translational-research-driving-seed-technology: intechopen.com, Retrieved 11 March, 2019

Seed Ecology

The study of ecological strategies which are used by plants in order to ensure their reproduction through seeds is known as seed ecology. The plants which produce seeds are known as spermatophytes. These are further sub-divided into two broad categories, namely, gymnosperms and angiosperms. This chapter has been carefully written to provide an easy understanding of these varied elements of study within seed ecology.

SPERMATOPHYTE

The spermatophytes, which means "seed plants", are some of the most important organisms on Earth. Life on land as we know it is shaped largely by the activities of seed plants. Soils, forests, and food are three of the most apparent products of this group. Seed-producing plants are probably the most familiar plants to most people, unlike mosses, liverworts, horsetails, and most other seedless plants which are overlooked because of their size or inconspicuous appearance. Many seed plants are large or showy. Conifers are seed plants; they include pines, firs, yew, redwood, and many other large trees. The other major group of seed-plants are the flowering plants, including plants whose flowers are showy, but also many plants with reduced flowers, such as the oaks, grasses, and palms.

Spermatophytes are divided into gymnosperms and angiosperms. The name Angiosperms comes from the Greek words: angeion, "vase", and sperm, "seeds". This name means that the seeds of these plants are not "naked", but encased in a specific structure, the ovary, that protects them from the external environment. The differences between Angiosperms and Gymnosperms are generally very clear.

The most important ones are:

- the Angiosperms include many wood or shrub species as well as many herbaceous species (unlike Gymnosperms);

- the cotyledons (i.e. the apparatuses that store food to feed the plant embryo) in Gymnosperms and few in Angiosperms (normally, one in monocotyledons sand two in dicotyledons);

- finally, in Gymnosperms the leaves are generally small and thin (needle- or scale-shaped leaves), while in Angiosperm they are more or less large and generally have an expanded lamina and reticulated or parallel ribs.

GYMNOSPERM

The gymnosperms, also known as Acrogymnospermae, are a group of seed-producing plants that includes conifers, cycads, Gin*kgo*, and gnetophytes. The name is based on the unenclosed condition of their seeds (called ovules in their unfertilized state). The non-encased condition of their seeds contrasts with the seeds and ovules of flowering plants (angiosperms), which are enclosed within an ovary. Gymnosperm seeds develop either on the surface of scales or leaves, which are often modified to form cones, or solitary as in yew, *Torreya, Ginkgo*.

The gymnosperms and angiosperms together compose the spermatophytes or seed plants. The gymnosperms are divided into six phyla. Organisms that belong to the Cycadophyta, Ginkgophyta, Gnetophyta, and Pinophyta (also known as Coniferophyta) phyla are still in existence while those in the Pteridospermales and Cordaitales phyla are now extinct.

By far the largest group of living gymnosperms are the conifers (pines, cypresses, and relatives), followed by cycads, gnetophytes (*Gnetum, Ephedra* and *Welwitschia*), and *Ginkgo biloba* (a single living species).

Roots in some genera have fungal association with roots in the form of mycorrhiza (*Pinus*), while in some others (*Cycas*) small specialised roots called coralloid roots are associated with nitrogen-fixing cyanobacteria.

Classification

The current formal classification of the living gymnosperms is the "Acrogymnospermae", which form a monophyletic group within the spermatophytes. The wider "Gymnospermae" group includes extinct gymnosperms and is thought to be paraphyletic. The fossil record of gymnosperms includes many distinctive taxa that do not belong to the four modern groups, including seed-bearing trees that have a somewhat fern-like vegetative morphology (the so-called "seed ferns" or pteridosperms). When fossil gymnosperms such as these and the Bennettitales, glossopterids, and *Caytonia* are considered, it is clear that angiosperms are nested within a larger gymnospermae clade, although which group of gymnosperms is their closest relative remains unclear.

The extant gymnosperms include 12 main families and 83 genera which contain more than 1000 known species.

Subclass Cycadidae

- Order Cycadales

  - ◦ Family Cycadaceae: Cycas

  - ◦ Family Zamiaceae: Dioon, Bowenia, Macrozamia, Lepidozamia, Encephalartos, Stangeria, Ceratozamia, Microcycas, Zamia.

Subclass Ginkgoidae

- Order Ginkgoales

  - ◦ Family Ginkgoaceae: Ginkgo

Subclass Gnetidae

- Order Welwitschiales

  - ◦ Family Welwitschiaceae: Welwitschia

- Order Gnetales

  - ◦ Family Gnetaceae: Gnetum

- Order Ephedrales

  - ◦ Family Ephedraceae: Ephedra

Subclass Pinidae

- Order Pinales

  - ◦ Family Pinaceae: Cedrus, Pinus, Cathaya, Picea, Pseudotsuga, Larix, Pseudolarix, Tsuga, Nothotsuga, Keteleeria, Abies

- Order Araucariales

  - ◦ Family Araucariaceae: Araucaria, Wollemia, Agathis

  - ◦ Family Podocarpaceae: Phyllocladus, Lepidothamnus, Prumnopitys, Sundacarpus, Halocarpus, Parasitaxus, Lagarostrobos, Manoao, Saxegothaea, Microcachrys, Pherosphaera, Acmopyle, Dacrycarpus, Dacrydium, Falcatifolium, Retrophyllum, Nageia, Afrocarpus, Podocarpus

- Order Cupressales

  - ◦ Family Sciadopityaceae: Sciadopitys

  - ◦ Family Cupressaceae: Cunninghamia, Taiwania, Athrotaxis, Metasequoia, Sequoia, Sequoiadendron, Cryptomeria, Glyptostrobus, Taxodium, Papuacedrus, Austrocedrus, Libocedrus, Pilgerodendron, Widdringtonia,

      Diselma, Fitzroya, Callitris, Actinostrobus, Neocallitropsis, Thujopsis, Thuja, Fokienia, Chamaecyparis, Cupressus, Juniperus, Calocedrus, Tetraclinis, Platycladus, Microbiota

   ○ Family Taxaceae: Austrotaxus, Pseudotaxus, Taxus, Cephalotaxus, Amentotaxus, Torreya

Diversity and Origin

There are over 1000 living species of gymnosperm. It is widely accepted that the gymnosperms originated in the late Carboniferous period, replacing the lycopsid rainforests of the tropical region. This appears to have been the result of a whole genome duplication event around 319 million years ago. Early characteristics of seed plants were evident in fossil progymnosperms of the late Devonian period around 383 million years ago. It has been suggested that during the mid-Mesozoic era, pollination of some extinct groups of gymnosperms was by extinct species of scorpionflies that had specialized proboscis for feeding on pollination drops. The scorpionflies likely engaged in pollination mutualisms with gymnosperms, long before the similar and independent coevolution of nectar-feeding insects on angiosperms. Evidence has also been found that mid-Mesozoic gymnosperms were pollinated by Kalligrammatid lacewings, a now-extinct genus with members which (in an example of convergent evolution) resembled the modern butterflies that arose far later.

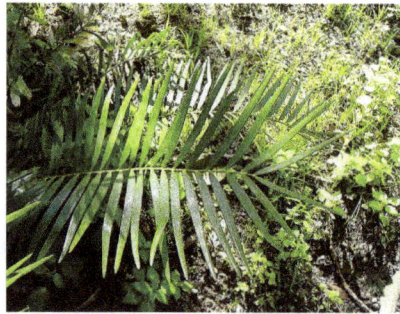

Zamia integrifolia, a cycad native to Florida.

Conifers are by far the most abundant extant group of gymnosperms with six to eight families, with a total of 65–70 genera and 600–630 species (696 accepted names). Conifers are woody plants and most are evergreens. The leaves of many conifers are long, thin and needle-like, other species, including most Cupressaceae and some Podocarpaceae, have flat, triangular scale-like leaves. *Agathis* in Araucariaceae and *Nageia* in Podocarpaceae have broad, flat strap-shaped leaves.

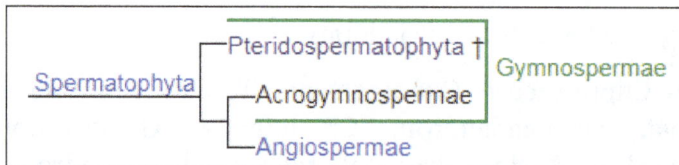

Cycads are the next most abundant group of gymnosperms, with two or three families, 11 genera, and approximately 338 species. A majority of cycads are native to tropical climates and are most abundantly found in regions near the equator. The other extant groups are the 95–100 species of Gnetales and one species of Ginkgo.

Uses

Gymnosperms have major economic uses. Pine, fir, spruce, and cedar are all examples of conifers that are used for lumber, paper production, and resin. Some other common uses for gymnosperms are soap, varnish, nail polish, food, gum, and perfumes.

Life Cycle

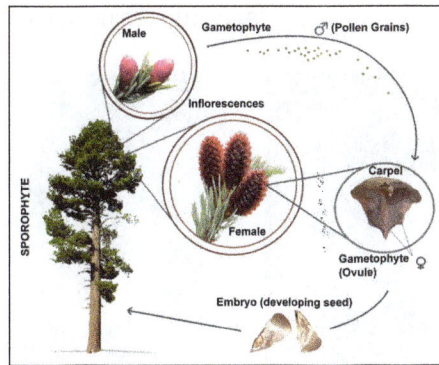

Example of gymnosperm lifecycle.

Gymnosperms, like all vascular plants, have a sporophyte-dominant life cycle, which means they spend most of their life cycle with diploid cells, while the gametophyte (gamete-bearing phase) is relatively short-lived. Two spore types, microspores and megaspores, are typically produced in pollen cones or ovulate cones, respectively. Gametophytes, as with all heterosporous plants, develop within the spore wall. Pollen grains (microgametophytes) mature from microspores, and ultimately produce sperm cells. Megagametophytes develop from megaspores and are retained within the ovule. Gymnosperms produce multiple archegonia, which produce the female gamete. During pollination, pollen grains are physically transferred between plants from the pollen cone to the ovule. Pollen is usually moved by wind or insects. Whole grains enter each ovule through a microscopic gap in the ovule coat (integument) called the micropyle.

The pollen grains mature further inside the ovule and produce sperm cells. Two main modes of fertilization are found in gymnosperms. Cycads and *Ginkgo* have motile sperm that swim directly to the egg inside the ovule, whereas conifers and gnetophytes have sperm with no flagella that are moved along a pollen tube to the egg. After syngamy (joining of the sperm and egg cell), the zygote develops into an embryo (young sporophyte). More than one embryo is usually initiated in each gymnosperm seed. The mature seed comprises the embryo and the remains of the female gametophyte, which serves as a food supply, and the seed coat.

ANGIOSPERM

Angiosperm is any of about 300,000 species of flowering plants, the largest and most diverse group within the kingdom Plantae. Angiosperms represent approximately 80 percent of all the known green plants now living. The angiosperms are vascular seed plants in which the ovule (egg) is fertilized and develops into a seed in an enclosed hollow ovary. The ovary itself is usually enclosed in a flower, that part of the angiospermous plant that contains the male or female reproductive organs or both. Fruits are derived from the maturing floral organs of the angiospermous plant and are therefore characteristic of angiosperms. By contrast, in gymnosperms (e.g., conifers and cycads), the other large group of vascular seed plants, the seeds do not develop enclosed within an ovary but are usually borne exposed on the surfaces of reproductive structures, such as cones.

Magnolia (Magnolia fraseri)

Orchids

Unlike such nonvascular plants as the bryophytes, in which all cells in the plant body participate in every function necessary to support, nourish, and extend the plant body (e.g., nutrition, photosynthesis, and cell division), angiosperms have evolved specialized cells and tissues that carry out these functions and have further evolved specialized vascular tissues (xylem and phloem) that translocate the water and nutrients to all areas of the plant body. The specialization of the plant body, which has evolved as an adaptation to a principally terrestrial habitat, includes extensive root systems that anchor the plant and absorb water and minerals from the soil; a stem that supports the growing plant body; and leaves, which are the principal sites of photosynthesis for most angiospermous plants. Another significant evolutionary advancement over the nonvascular and the more primitive vascular plants is the presence of localized regions for plant growth, called meristems and cambia, which extend the length and width of the plant body, respectively. Except under certain conditions, these regions are the only areas in which mitotic cell division takes place in the plant body, although cell differentiation continues to occur over the life of the plant.

The angiosperms dominate Earth's surface and vegetation in more environments, particularly terrestrial habitats, than any other group of plants. As a result, angiosperms are the most important ultimate source of food for birds and mammals, including humans. In addition, the flowering plants are the most economically important group of green plants, serving as a source of pharmaceuticals, fibre products, timber, ornamentals, and other commercial products.

Quinoa plant Quinoa (Chenopodium quinoa) growing in the Bolivian Altiplano region.

Although the taxonomy of the angiosperms is still incompletely known, the latest classification system incorporates a large body of comparative data derived from studies of DNA sequences. It is known as the Angiosperm Phylogeny Group IV (APG IV) botanical classification system. The angiosperms came to be considered a group at the division level (comparable to the phylum level in animal classification systems) called Anthophyta, though the APG system recognizes only informal groups above the level of order.

Honeysuckle A yellow-orange honeysuckle.

Features

The variety of forms found among angiosperms is greater than that of any other plant group. The size range alone is quite remarkable, from the smallest individual flowering plant, probably the watermeal (Wolffia; Araceae) at less than 2 millimetres (0.08 inch), to one of the tallest angiosperms, Australia's mountain ash tree (Eucalyptus regnans; Myrtaceae) at about 100 metres (330 feet). Between these two extremes lie angiosperms of almost every size and shape. Examples of this variability include the succulent cacti

(Cactaceae), the fragile orchids (Orchidaceae), the baobabs (Adansonia species; Malvaceae), vines, rosette plants such as the dandelion (Asteraceae), and carnivorous plants such as sundews (Drosera; Droseraceae) and the Venus flytrap (Dionaea muscipula; Droseraceae). To understand this vast array of forms, it is necessary to consider the basic structural plan of the angiosperms.

Orchid (Vanda).

Dandelion (Taraxacum officinale).

The basic angiosperm form is woody or herbaceous. Woody forms (generally trees and shrubs) are rich in secondary tissues, while herbaceous forms (herbs) rarely have any. Annuals are herbs that complete their growing cycle (growth, flowering, and death) within the same season. Examples of annuals can be found among cultivated garden plants, such as beans (Phaseolus and other genera; Fabaceae), corn (maize, Zea mays; Poaceae), and squashes (Cucurbita; Cucurbitaceae), as well as among the wildflowers, such as some buttercups (Ranunculus; Ranunculaceae) and poppies (Papaver and other genera; Papaveraceae). Biennials are also herbs, but, unlike annuals, their growing cycle spans two years: the vegetative (nonreproductive) plant growth takes place from seed during the first year, and flowers and fruit develop during the second. The beet (Beta vulgaris; Amaranthaceae) and carrot (Daucus carota; Apiaceae) are well-known biennials.

Cherry trees blossoming Cherry trees blossoming in spring at an orchard.

A perennial grows for many years and often flowers annually. In temperate areas the aerial parts of a perennial die back to the ground at the end of each growing season and new shoots are produced the following season from such subterranean parts as bulbs, rhizomes, corms, tubers, and stolons.

Yellow poplar (Liriodendron tulipifera).

Active traps of the Venus's-flytrap (Dionaea muscipula), a carnivorous plant. If depressed at least twice, thin pressure-sensitive hairs in the trap stimulate the lobes to clamp tightly over an insect.

Structure

The basic angiosperm body has three parts: roots, stems, and leaves. These primary organs constitute the vegetative (nonreproductive) plant body. Together, the stem and its attached leaves constitute the shoot. Collectively, the roots of an individual plant make up the root system and the shoots the shoot system.

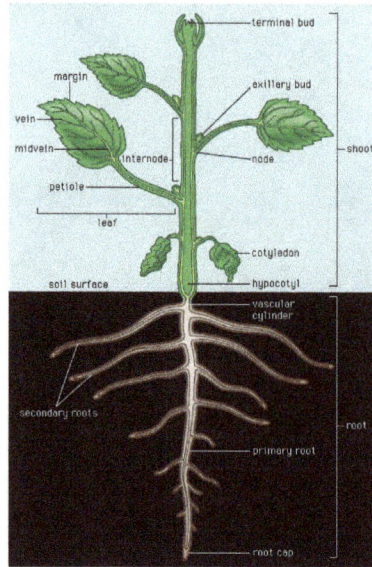

A typical dicotyledonous plant.

Root Systems

The roots anchor a plant, absorb water and minerals, and provide a storage area for food. The two basic types of root systems are a primary root system and an adventitious root system. The most common type, the primary system, consists of a taproot (primary root) that grows vertically downward (positive geotropism). From the taproot are produced smaller lateral roots (secondary roots) that grow horizontally or diagonally. These secondary roots further produce their own smaller lateral roots (tertiary roots). Thus, many orders of roots of descending size are produced from a single prominent root, the taproot. Most dicotyledons produce taproots, as, for example, the dandelion (Taraxacum officinale).

In some cases, the taproot system is modified into a fibrous, or diffuse, system, in which the initial secondary roots soon equal or exceed the primary root in size. The result is several large, positively geotropic roots that produce higher-order roots, which may also grow to the same size. Thus, in fibrous root systems there is no well-defined single taproot. In general, fibrous root systems are shallower than taproot systems.

The second type of root system, the adventitious root system, differs from the primary variety in that the primary root is often short-lived and is replaced or supplemented by many roots that form from the stem. Most monocotyledons have adventitious roots; examples include orchids (Orchidaceae), bromeliads (Bromeliaceae), and many other epiphytic plants in the tropics. Grasses (Poaceae) and many other monocotyledons produce fibrous root systems with the development of adventitious roots.

Adventitious roots, when modified for aerial support, are called prop roots, as in corn or some figs (Ficus; Moraceae). In many tropical rainforest trees, large woody prop roots develop from adventitious roots on horizontal branches and provide additional anchorage and support. Many bulbous plants have contractile adventitious roots that pull the bulb deeper into the ground as it grows. Climbing plants often grip their supports with specialized adventitious roots. Some lateral roots of mangroves become specialized as pneumatophores in saline mud flats; pneumatophores are lateral roots that grow upward (negative geotropism) for varying distances and function as the site of oxygen intake for the submerged primary root system. The plants mentioned are only a few examples of root diversity in angiosperms, a condition that is unparalleled in any other vascular plant group.

Pneumatophores of the black mangrove (Avicennia germinans) encrusted with salt and a young seedling projecting above the surface of the water.

Many primary root and adventitious root systems have become modified for special functions, the most common being the formation of tuberous (fleshy) roots for food storage. For example, carrots and beets are tuberous roots that are modified from taproots, and cassava (manioc) is a tuberous root that is modified from an adventitious root.

Stems

The stem is an aerial axis of the plant that bears leaves and flowers and conducts water and minerals from the roots and food from the site of synthesis to areas where it is to be used. The main stem of a plant is continuous with the root system through a transition region called the hypocotyl. In the developing embryo, the hypocotyl is the embryonic axis that bears the seedling leaves (cotyledons).

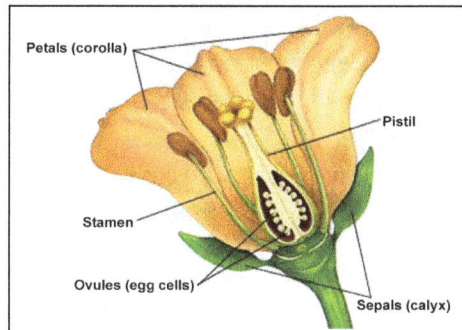

Parts of a flower.

In a maturing stem, the area where a leaf attaches to the stem is called a node, and the region between successive nodes is called an internode. Stems bear leafy shoots (branches) at the nodes, which arise from buds (dormant shoots). Lateral branches develop either from axillary, or lateral, buds found in the angle between the leaf and the stem or from terminal buds at the end of the shoot. In temperate-climate plants these buds have extended periods of dormancy, whereas in tropical plants the period of dormancy is either very short or nonexistent.

The precise positional relationship of stem, leaf, and axillary bud is important to understanding the diversity of the shoot system in angiosperms. Understanding this relationship makes it possible to identify organs such as leaves that are so highly modified they no longer look like leaves, or stems that are so modified that they resemble leaves.

Branching in angiosperms may be dichotomous or axillary. In dichotomous branching, the branches form as a result of an equal division of a terminal bud (i.e., a bud formed at the apex of a stem) into two equal branches that are not derived from axillary buds, although axillary buds are present elsewhere on the plant body. The few examples of dichotomous branching among angiosperms are found only in some cacti, palms (Arecaceae), and bird-of-paradise plants (Strelitziaceae).

Traveler's trees (Ravenala madagascariensis).

The two modes of axillary branching in angiosperms are monopodial and sympodial. Monopodial branching occurs when the terminal bud continues to grow as a central leader shoot and the lateral branches remain subordinate—e.g., beech trees (Fagus; Fagaceae). Sympodial branching occurs when the terminal bud ceases to grow (usually because a terminal flower has formed) and an axillary bud or buds become new leader shoots, called renewal shoots—e.g., the Joshua tree (Yucca brevifolia; Asparagaceae). Plants with monopodial growth are usually pyramidal in overall shape, while those with sympodial growth often resemble a candelabra.

By combining monopodial and sympodial branching in one plant, many different tree architectures have evolved. A simple example is found in dogwoods (Cornus; Cornaceae), where the main axis is monopodial and the lateral branches are sympodial.

leaves; beech Sunshine on the leaves of a beech tree (Fagus).

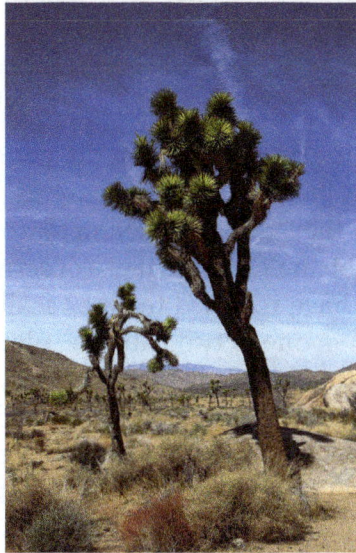

Joshua tree Joshua trees (Yucca brevifolia).

Very different plant forms result from simply changing the lengths of the internodes. Extreme shortening of the internodes results in rosette plants, such as lettuce (Lactuca sativa; Asteraceae), in which the leaves develop but the internodes between them do not elongate until the plant "bolts" when flowering. Extreme lengthening of the internodes often results in twining vines, as in the yam (Dioscorea esculenta; Dioscoreaceae).

Leaves

The basic angiosperm leaf is composed of a leaf base, two stipules, a petiole, and a blade (lamina). The leaf base is the slightly expanded area where the leaf attaches to the stem. The paired stipules, when present, are located on each side of the leaf base and may resemble scales, spines, glands, or leaflike structures. The petiole is a stalk that connects the blade with the leaf base. The blade is the major photosynthetic surface of the plant and appears green and flattened in a plane perpendicular to the stem.

Stipules of the tulip tree (Liriodendron tulipifera). Stipules develop
at the base of a leaf and protect the developing blade.

When only a single blade is inserted directly on the petiole, the leaf is called simple. Simple leaves may be variously lobed along their margins. The margins of simple leaves may be entire and smooth or they may be lobed in various ways. The coarse teeth of dentate margins project at right angles, while those of serrate margins point toward the leaf apex. Crenulate margins have rounded teeth or scalloped margins. Leaf margins of simple leaves may be lobed in one of two patterns, pinnate or palmate. In pinnately lobed margins the leaf blade (lamina) is indented equally deep along each side of the midrib (as in the white oak, Quercus alba; Fagaceae), and in palmately lobed margins the lamina is indented along several major veins (as in the red maple, Acer rubrum; Sapindaceae). A great variety of base and apex shapes also are found.

Common leaf morphologies.

Many leaves contain only some of these leaf parts; for example, many leaves lack a petiole and so are attached directly to the stem (sessile), and others lack stipules (ex-stipulate). In compound leaves, a blade has two or more subunits called leaflets: in palmately compound leaves, the leaflets radiate from a single point at the distal end of the petiole; in pinnately compound leaves, a row of leaflets forms on either side of an extension of the petiole called the rachis. Some pinnately compound leaves branch again, developing a second set of pinnately compound leaflets (bipinnately compound). The many degrees of compoundness in highly elaborated leaves, such as bipinnately or tripinnately compound, cause these leaves to often appear to be shoot systems. It is always possible to distinguish them, however, because axillary buds are found in the angle between the stem and the petiole (axil) of pinnately or palmately compound leaves but not in the axils of leaflets.

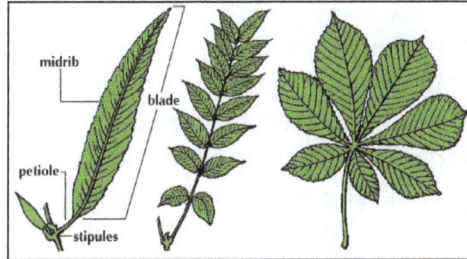

The willow leaf (left) is simple. The walnut leaf (middle) is pinnately compound, or feather-shaped. The horse chestnut leaf (right) is palmately compound, or hand-shaped.

The three patterns of leaf arrangement on stems in angiosperms are alternate, opposite (paired), and whorled. In alternate-leaved plants, the leaves are single at each node and borne along the stem alternately in an ascending spiral. In opposite-leaved plants, the leaves are paired at a node and borne opposite to each other. A plant has whorled leaves when there are three or more equally spaced leaves at a node.

Basil (Ocimum basilicum). The simple leaves are arranged oppositely along the stems.

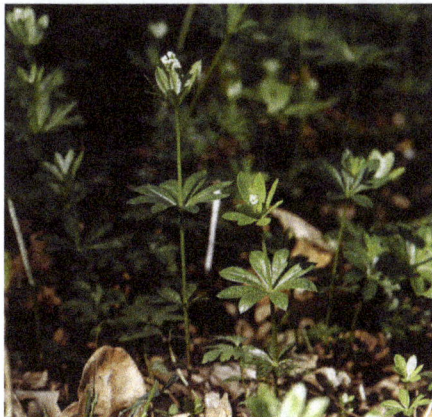

sweet woodruff The whorled leaves of sweet woodruff.

Leaf Modifications

Whole leaves or parts of leaves are often modified for special functions, such as for climbing and substrate attachment, storage, protection against predation or climatic conditions, or trapping and digesting insect prey. In temperate trees, leaves are simply protective bud scales; in the spring when shoot growth is resumed, they often exhibit a complete growth series from bud scales to fully developed leaves.

Stipules often develop before the rest of the leaf; they protect the young blade and then are often shed when the leaf matures. Spines are also modified leaves. In cacti, spines are wholly transformed leaves that protect the plant from herbivores, radiate heat from the stem during the day, and collect and drip condensed water vapour during the cooler night. In the many species of the spurge family (Euphorbiaceae), the stipules are modified into paired stipular spines and the blade develops fully. In ocotillo (Fouquieria splendens; Fouquieriaceae), the blade falls off and the petiole remains as a spine.

Euphorbia fianarantsoae.

Many desert plants, such as stoneplants (Lithops; Aizoaceae) and aloe (Aloe; Asphodelaceae), develop succulent leaves for water storage. The most common form of storage leaves are the succulent leaf bases of underground bulbs (e.g., tulip and Crocus) that serve as either water- or food-storage organs or both. Many nonparasitic plants that grow on the surfaces of other plants (epiphytes), such as some of the bromeliads, absorb water through specialized hairs on the surfaces of their leaves. In the water hyacinth (Eichhornia crassipes), swollen petioles keep the plant afloat.

Lithops

Water hyacinth (Eichhornia crassipes).

Leaves or leaf parts may be modified to provide support. Tendrils and hooks are the most common of these modifications. In the flame lily (Gloriosa superba; Colchicaceae), the leaf tip of the blade elongates into a tendril and twines around other plants for support. In the garden pea (Pisum sativum; Fabaceae), the terminal leaflet of the compound leaf develops as a tendril. In nasturtium (Tropaeolum majus; Tropaeolaceae) and Clematis (Ranunculaceae), the petioles coil around other plants for support. In catbrier (Smilax; Smilacaceae), the stipules function as tendrils. In certain vining angiosperms with compound leaves, some of the leaflets have modified into grapnel-like hooks—e.g., Tecoma radicans. Many monocotyledons have sheathing leaf bases that are concentrically arranged and form a pseudotrunk, as in banana (Musa). In many epiphytic bromeliads, the pseudotrunk also functions as a water reservoir.

Tendrils of catbrier (Smilax rotundifolia).

Carnivorous plants use their leaves to attract and trap insects. Glands in the leaves secrete enzymes that digest the captured insects, and the leaves then absorb the nitrogenous compounds (amino acids) and other products of digestion. Plants that use insects as a nitrogen source tend to grow in nitrogen-deficient soils.

Slender pitcher plant Pitcher-shaped leaves of the carnivorous
slender pitcher plant (Nepenthes gracilis).

Shoot System Modifications

Entire shoot systems are often modified for such special functions as climbing, protection, adaptation to arid habitats, and water or food storage. The modifications generally involve structural and shape changes to the stem and the reduction of the leaves to small scales. Many of the modifications parallel those previously described for leaves. In the passion flower (Passiflora; Passifloraceae) and grape (Vitis vinifera; Vitaceae), axillary buds develop as tendrils with reduced leaves and suppressed axillary buds. In the grape these axillary tendrils are actually modified and reduced inflorescences. In the plant from which strychnine is obtained (Strychnos nux-vomica), the axillary buds develop into hooks for climbing.

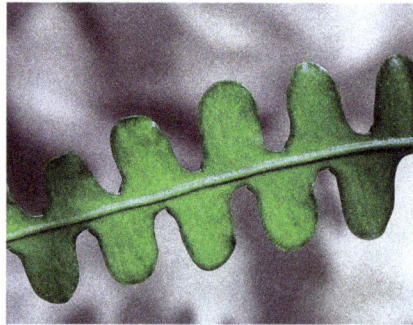

A cladode of the orchid, or leaf, cactus (Epiphyllum). The stem does not bear leaves
but rather becomes flattened and leaflike, assuming the plant's photosynthetic functions.

Thorns represent the modification of an axillary shoot system in which the leaves are reduced and die quickly and the stems are heavily sclerified and grow for only a limited time (determinate growth). Thorns appear to protect the plant against herbivores. Examples are found in the Bougainvillea (Nyctaginaceae), where the thorn is a modified inflorescence, the honey locust (Gleditsia triacanthos; Fabaceae), the anchor plant (Colletia paradoxa; Rhamnaceae), and Citrus (Rutaceae).

Honey locust trunk Thorny trunk of the common honey locust (Gleditsia triacanthos).

Cladodes (also called cladophylls or phylloclades) are shoot systems in which leaves do not develop; rather, the stems become flattened and assume the photosynthetic functions of the plant. In asparagus (Asparagus officinalis; Asparagaceae), the scales found on the asparagus spears are the true leaves. If the thick, fleshy asparagus spears continue to grow, flat, green, leaflike structures called cladodes develop in the axils of the scale leaves. The presence of cladodes in unrelated desert angiosperm families is an excellent example of convergent evolution, or the independent development of the same characteristic in unrelated taxa.

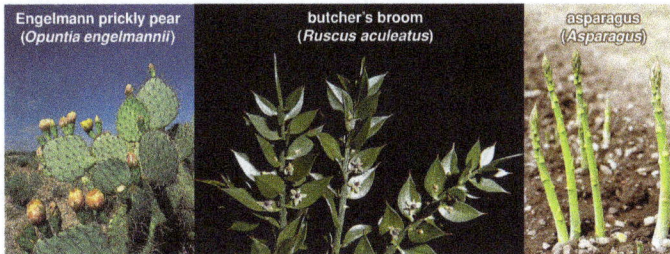

Cladode Examples of plants with cladodes: (left) Engelmann prickly pear
(Opuntia engelmannii), (centre) butcher's broom (Ruscus aculeatus),
and (right) asparagus (Asparagus) shoots.

All cacti (Cactaceae) have cladodes, and many desert members of the spurge (Euphorbiaceae) and milkweed (Apocynaceae) families have similar vegetative morphologies that are derived by modifying different parts to look and function in the same way. Each of these plant groups has columnar, water-storing green stems, reduced leaves, and protective spines or thorns. They are often called stem succulents. In the cacti, the leaves on the main stems last for a very short time (they do not even develop as scale leaves) and the leaves of the axillary buds (the round cushion areas, or areoles, on the trunks) develop as spines. In many members of the Euphorbiaceae, the leaves on the main stems are green but short-lived, and the stipules develop as spines. In a number of plants of the Apocynaceae family, the leaves are also small and ephemeral, and the axillary buds develop as thorns. The cacti are almost exclusively New World plants adapted to dry or arid habitats, and the Euphorbiaceae and Apocynaceae occur in similar habitats in Asia and Africa. The reduction of leaves is so extreme in the Cactaceae that the epiphytic cacti (e.g., Epiphyllum) of the Neotropics can no longer produce leaves; rather, they produce thin, flat cladodes that superficially resemble leaves.

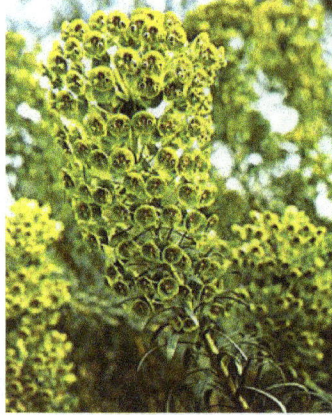
Spurge (Euphorbia venata).

Many shoot systems have been modified into organs of food storage, reproduction, or both, called rhizomes, tubers, and corms. Rhizomes are distinguished from roots in having nodes with reduced leaves and internodes. Rhizomes are horizontal, usually subterranean shoots with scale leaves and adventitious roots on the underside. Their chief functions are vegetative reproduction and food storage; food stored in the rhizomes allows these plants to survive drought and extended winters. Most rhizomes are perennial, sending up new shoots from the nodes and spreading the colony. Often the terminal bud of a rhizome becomes upright and then flowers, with a rhizome axillary bud becoming a renewal shoot. Many economically important plants, such as banana, and almost all grasses, including bamboo, and sugarcane, have rhizomes. Such plants are propagated primarily by fragmentation of the rhizome. In some plants, the growing tips of rhizomes become much enlarged food storage organs called tubers. The common potato (Solanum tuberosum; Solanaceae) forms such tubers. The much-reduced scale leaves and their associated axillary buds form the eyes of the potato. Tubers should not be confused with tuberous roots. Tubers are modified shoots, whereas tuberous roots are modified roots. The common feature, and hence the similar names, derives from the fleshy nature of both organs. Tubers and tuberous roots function in water and food storage, but only tubers are involved in vegetative (nonsexual) reproduction. Tuberous roots develop from taproots in carrots and from adventitious roots in dahlias (Dahlia; Asteraceae).

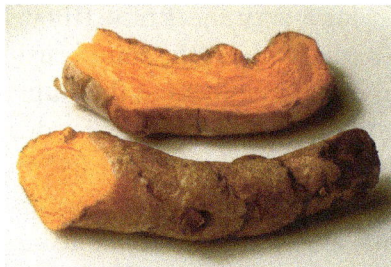
Turmeric Rhizomes of common turmeric (Curcuma longa).

Underground stems examples of underground stems: (left) taro corm,
(centre) ginger rhizome, and (right) potato tubers.

Another distinctive modification for food storage is the corm, a short, upright shoot system with a thick, hard stem covered with thin membranous scale leaves as in jack-in-the-pulpit (Arisaema triphyllum; Araceae) and gladiolus (Gladiolus; Iridaceae). Corms are usually hard and fibrous and function for overwintering and drought resistance.

Slender creeping stems that grow above the soil surface are called stolons, or runners. Stolons have scale leaves and can develop roots and, therefore, new plants, either terminally or at a node. In the strawberry (Fragaria; Rosaceae), the stolons are used for propagation: buds appear at nodes along the stolons and develop into new strawberry plants.

Distribution and Abundance

The diversity of form within the angiosperms has contributed to their successful colonization of more habitats than any other group of land plants. Gymnosperms (the non-flowering seed plants) are only woody plants with a few woody twining vines. There are few herbaceous or aquatic gymnosperms; most gymnosperms do not occur as swampy vegetation or in marine habitats. With the exception of cycads, gymnosperms have simple leaves, and none are modified as spines, tendrils, or storage organs.

The absence of substantial diversity in the vegetative features of gymnosperms appears to have limited their ability to adapt to diverse or extreme habitats. The absence of xylem vessels in most gymnosperms, and hence the less efficient water transport system than that found in the angiosperms, is one example. In fact, the only gymnosperms with vessels, the Gnetales, is the only group that contains vines and the only group that deviates from the usually woody trunk growth form. The absence of vessels in angiosperms, however, is rare; the few groups without vessels are small trees or shrubs with limited distribution, as in the Winteraceae. Another factor contributing to the limited distribution of gymnosperms is that they do not produce reproductive structures until several years after the seed germinates; therefore, a woody habit is required to achieve sexual maturity. Finally, the gymnosperms also require a relatively stable environment for growth. Thus, restraints imposed by anatomy and life cycle have probably limited morphological diversity among the gymnosperms.

The wide variation in the angiosperm form is reflected in the range of habitats in which they grow and their almost complete worldwide distribution. The only area without angiosperms is the southern region of the Antarctic continent, although two angiosperm groups are found in the islands off that continent. Angiosperms dominate terrestrial vegetation, particularly in the tropics, although submerged and floating aquatic angiosperms do exist throughout the world. Angiosperms are the principal component of salt marshes, tidal marshes, and mangrove marshes. The only vascular marine plants are a few submerged marine angiosperms that occur in shallow waters of coastal areas throughout the world—for example, the eelgrasses (Zostera and Phyllospadix; Zosteraceae). The various terrestrial biomes (defined primarily based upon the type of vegetation and climate) are composed mainly of herbaceous and woody angiosperms, except for taiga (boreal forest), temperate rainforest, and juniper savanna, where conifers (a gymnospermous division) dominate the woody component and angiosperms dominate the herbaceous and shrub components.

Common eelgrass Common eelgrass (Zostera marina). A marine flowering plant, common eelgrass can be found in cooler coastal waters.

Morphological and habitat diversity, together with cosmopolitan distribution, contributes to the wide ecological tolerance of the angiosperms—adapting to Alpine tundra regions and salt marshes, from the Arctic Circle to the lowland tropical rainforests. The importance of angiosperms in the terrestrial portion of the biosphere is rarely rivaled by any other group of organisms.

Indian pipe (Monotropa uniflora).

All but a few angiosperms are autotrophs: they are green plants (primary producers) that use solar radiation, carbon dioxide, water, and minerals to synthesize organic compounds; oxygen is a by-product of these metabolic reactions. The few exceptions are either mycoheterotrophs (e.g., the Indian pipe Monotropa uniflora; Ericaceae) that use connections with mycorrhizal fungi (fungi that form an association with the roots of certain plants) to obtain carbohydrates or parasitic plants that develop specialized roots (haustoria), which penetrate the host plant and absorb food and other materials (e.g., the dodder).

Importance

Contribution to Food Chain

Because angiosperms are the most numerous component of the terrestrial environment in terms of biomass and number of individuals, they provide an important source of food for animals and other living organisms. Organic compounds (carbon-containing compounds, principally carbohydrates) not only are used by the plant itself for synthesizing cellular structures and for fueling their basic metabolisms but also serve as the only source of energy for most heterotrophic organisms. (Heterotrophs require an organic source of carbon that has originated as part of another living organism, in contrast to autotrophs, which require only an inorganic source of carbon—CO_2.) Solar energy is trapped by the photosynthetic pigments in the plant cells and converted into chemical energy, which is stored in the tissues of the plant. The trapped energy is transferred from one organism to the next as herbivores consume the plant, carnivores consume herbivores, and so on up the food chain. In a temperate forest, a single angiosperm tree may support many thousands of animals (the majority being insects, birds, and mammals), a relationship that underscores the basic importance of the angiosperms to the food chain and the ecological web.

The angiosperm body contributes to the food chain in many ways. The vegetative parts (the nonreproductive organs, such as stems and leaves) are consumed by, and support, plant-eating animals. Vast numbers of insects and other invertebrates depend on shoots for food during all or part of their life histories. The reproductive organs (flowers, fruits, and seeds) also provide an energy source for many animals. The pollen supports many pollinating insects, particularly bees.

The flowers provide food from floral nectaries that secrete sugars and amino acids. These flowers often produce fragrances that attract pollinators which feed on the nectar. Nectar-feeding animals include many insect groups (bees, butterflies, moths, flies, and even mosquitoes), many mammal groups (bats, small rodents, and small marsupials), and birds (honeyeaters, hummingbirds, and sunbirds). Nectaries also occur on the nonfloral, or vegetative, parts of some angiosperms, such as the leaves and the petioles of bull's-horn thorn (Acacia collinsii; Fabaceae). Ants live inside the hollow modified spinous structures of bull's-horn thorn and feed on the nectar. In return for this food source, they attack and destroy animals of all sizes as well as other plants that contact the acacia plant. In doing so, the ants protect the bull's-horn thorn from herbivores and other plants competing for the available space, light, and minerals.

A honeybee (Apis mellifera) pollinating a blue iris (Iris). Flecks of pollen grains dislodged from the stamens by the foraging bee can be seen on the bee's body.

Fruits produced by angiosperms are the principal food for many bats, birds, mammals, and even some fish. Seeds are also an important food source for many animals, particularly small rodents and birds. These animals often carry the fruits and seeds of the angiosperms they consume to new areas, where the angiosperms propagate.

Another aspect of angiosperm diversity is found in the production of secondary compounds, such as alkaloids, quinones, essential oils, and glycosides. Angiosperms have evolved a comprehensive array of unpalatable or toxic secondary plant compounds that protect the plants from foraging herbivores. Some insects, however, successfully store these secondary compounds in their tissues and use them as protection from predation.

As the principal component of the terrestrial biosphere, the angiosperm flora determines many features of the habitat, some of which are available food, aspects of the forest canopy, and grazing land. They supply nesting sites and materials for a wide range of birds and mammals, and they are the principal living spaces for many primates, reptiles, and amphibians. The tank bromeliad, which traps water in its crowns, provides a habitat for salamanders, frogs, and many aquatic insects and larvae. The animal inhabitants of the water-filled insectivorous pitcher plant leaves have adapted to the hostile environment of the leaves' digestive fluids.

Significance to Humans

Angiosperms are as important to humans as they are to other animals. Angiosperms serve as the major source of food—either directly or indirectly through consumption by herbivores—and, they are a primary source of consumer goods, such as building materials, textile fibres, spices and herbs, and pharmaceuticals.

Potato (Solanum tuberosum).

Among the most important food plants on a global scale are cereals from the grass family (Poaceae); potatoes, tomatoes, eggplant, and chili peppers from the potato family (Solanaceae); legumes or beans (Fabaceae); pumpkins, melons, and gourds from the squash family (Cucurbitaceae); broccoli, cabbage, cauliflower, radish, and other vegetables from the mustard family (Brassicaceae); and almonds, apples, apricots, cherries, loquats, peaches, pears, raspberries, and strawberries from the rose family (Rosaceae). Members of many angiosperm families are used for food on a local level, such as ullucu (Ullucus tuberosus; Basellaceae) in the Andes and cassava (Manihot esculenta; Euphorbiaceae) throughout the tropics. Tropical angiosperm trees are an important source of timber in the tropics and throughout the world.

The flowering plants have a number of uses as food, specifically as grains, sugars, vegetables, fruits, oils, nuts, and spices. In addition, plants and their products serve a number of other needs, such as dyes, fibres, timber, fuel, medicines, and ornamentals. Many plants serve more than one function. For example, the seeds of the kapok fruit (Ceiba pentandra; Malvaceae) yield a water-repellent fibre used in sound and thermal insulation and an edible oil used in cooking, lubricants, and soap; the oil cake is rich in protein and is fed to livestock; and the soft, light wood is used to make furniture and boats.

Woolly seeds produced by the seed pods of the kapok tree (Ceiba pentandra).

The angiospermous plant converts the energy of the sun into starch, the energy-rich storage form of sugar, and reserves it in the endosperm of the seed for the time when the seedling germinates and grows. Among the most economically important grains throughout the world are corn (Zea mays), wheat (Triticum), rice (Oryza), barley (Hordeum), oats (Avena), sorghum (Sorghum), and rye (Secale), all members of the grass family, Poaceae.

Corn provides food for humans and domesticated animals, and its derivatives (e.g., cornstarch and corn oil) are used in making cosmetics, adhesives, varnishes, paints, soaps, and linoleum. Among the many cultivars of Zea mays, dent corn, variety indentat, is a widely used feed type in the United States. Wheat, barley, and rye are all members of the same tribe (Triticeae) within the family Poaceae. Wheat is among the oldest of the cultivated food crops. Barley is used for human consumption, livestock feed, and malting. Rye is usually used as a livestock feed, but can be used in baking and distilling liquor. Rice is a semiaquatic annual grass and is one of the major cereal crops of the world.

Vegetables constitute perhaps the greatest diversity of form and nutritional content and are grown for one or more of their parts—the flowers, shoots, or leaves; or the underground parts, such as tuberous roots, bulbs, rhizomes, corms, and tubers.

The globe, or French, artichoke (Cynara scolymus; Asteraceae) is an immature flower bud and receptacle overlaid by bracts. Asparagus (Asparagus officinalis; Asparagaceae) is a perennial plant cultivated for its succulent green cladodes that arise from underground stems called crowns.

The mustard family (Brassicaceae) contains a number of important vegetables—broccoli, Brussels sprouts, cabbage, cauliflower, collards, kale, and kohlrabi—all members of Brassica oleraceae and comprising a group of vegetables called the cole crops, a term that probably reflects the fact that they are principally stem plants. The flower heads and stalks of broccoli and cauliflower are eaten, the two plants differing in that the white head of the cauliflower consists of malformed (hypertrophied) flowers that form in dense clusters. Brussels sprouts continually form many small heads in the axils of the leaves throughout the growing season. The cabbage head is a large terminal bud.

Salinas, California: mustard Field of mustard in flower.

The edible portion of celery (Apium graveolens; Apiaceae) is the petiole (leaf stalk) that arises from a compact stem. Rhubarb (Rheum rhabarbarum; Polygonaceae) is a leafy plant also grown for its leaf petioles.

Parsley (Petroselinum crispum; Apiaceae), spinach (Spinacia oleracea; Amaranthaceae), and Swiss chard (Beta vulgaris, variety cicla; Amaranthaceae) are cultivated for their leaves, and the leek (Allium porrum; Amaryllidaceae), a close relative of the onion, is cultivated for its leaf bases.

Root crops are grown for their fleshy subterranean storage bodies: tuberous roots, bulbs, rhizomes, corms, and tubers. The potato is a tuber found in Solanaceae, the potato family. Other important root crops include the carrot (Daucus carota; Apiaceae), beet (Beta vulgaris; Amaranthaceae), and sweet potato (Ipomoea batatas; Convolvulaceae), as well as the radish (Raphanus sativus), turnip (Brassica rapa, variety rapa), and rutabaga (B. napus, variety napobrassica) of the mustard family (Brassicaceae).

Sugar beet (Beta vulgaris).

Bulb crops are underground leafy scales attached to short compressed stems; food is stored in the leaves rather than the roots, causing them to enlarge into bulbs. Onions and garlic (Allium cepa and A. sativum, respectively; Amaryllidaceae) are the most obvious examples of the bulb vegetable.

Many plants classified popularly as vegetables are in actuality fruits because they develop from the reproductive structures of the plant. The genus Cucurbita (Cucurbitaceae) includes the pumpkins, squashes, and gourds, of which C. moschata (winter squash, or crookneck pumpkin), C. pepo (summer squash, or marrow), and C. mixta (the pumpkin, or mixta squash) are some of the common types. The cucumber (Cucumis sativus; Cucurbitaceae) produces a fruit that develops from a branching vine. Okra (Abelmoschus esculentus; Malvaceae) is a warm-weather crop that produces small fruit pods. Breadfruit (Artocarpus altilis; Moraceae), a plant native to the Pacific Islands, is a staple crop, providing a rich source of calcium and starch.

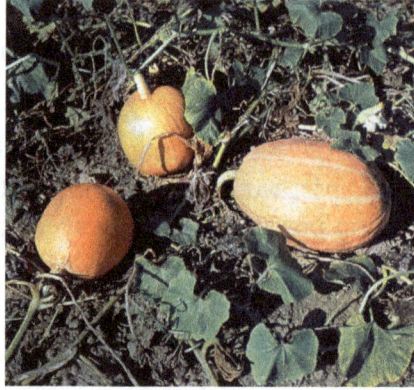

Pumpkin (Cucurbita).

The common bean (Phaseolus vulgaris), including the French, or kidney, bean, the string bean, and the navy bean, is the edible fleshy pod containing the bean seeds. It provides a good source of protein. Lima beans (P. lunatus) probably originated in Central America and are now found in the United States, the lowland tropics, and Africa. The garden, or English, pea (Pisum sativum; Fabaceae) is an annual cool-weather plant cultivated for its edible green seed or pod. The pea is found throughout most temperate and tropical regions.

The family Solanaceae contains a number of important fruit vegetables—eggplants (aubergines), peppers, and tomatoes—all herbaceous plants, which are perennial in the tropics and annual in temperate zones. The pepper (Capsicum) includes the sweet, or bell, pepper (which is green when immature but red or yellow when ripe), and the red, or chili, pepper. Pepper plants are cultivated for their fruits, some of which are extremely pungent because of the presence of capsaicin found in the septa, in the placenta, and, to a lesser extent, in the seeds, but not in the wall, of the fruit. The tomato (Solanum lycopersicum), native to South America, was at one time wrongly reported to bear poisonous fruits. The fruit is a fleshy berry invested with many small seeds.

Tomato (Solanum lycopersicum).

Plants cultivated for their fruits are found in temperate, tropical, or subtropical regions. Temperate plants are generally deciduous and either tolerate or require a cool period for growth. Apples (Malus) and pears (Pyrus) are important pome fruits of the family Rosaceae. Some well-known stone fruits of the family include the peaches and nectarines (Prunus persica), plums (P. domestica), and cherries (P. avium). Other temperate fruits grown on bushes, vines, or low plants include grapes (Vitis; Vitaceae) and strawberries (Fragaria; Rosaceae), as well as blueberries (Vaccinium) and cranberries (V. macrocarpon), both from Ericaceae.

Tropical fruits tend to be grown on evergreen plants and can survive temperatures only above freezing. Subtropical plants are either deciduous or tropical and are not as susceptible to temperatures slightly below freezing. Citrus (Rutaceae), avocados (Persea americana; Lauraceae), olives (Olea; Oleaceae), dates (Phoenix dactylifera; Arecaceae), figs (Ficus; Moraceae), pineapples (Ananas comosus; Bromeliaceae), bananas (Musa; Muscaceae), and papayas (Carica; Caricaceae) are tropical and subtropical plants.

Common fig Fruit of the common fig (Ficus carica).

Commercially important plants cultivated for the nuts and hard seeds they produce are almonds (Prunus dulcis; Rosaceae), walnuts (Juglans; Juglandaceae), pecans (Carya illinoinensis; Juglandaceae), macadamias (Macadamia; Proteaceae), and hazelnuts (Corylus; Betulaceae).

Macadamia (Macadamia ternifolia).

Sugarcane (Saccharum officinurum; Poaceae) and sugar beet (Beta vulgaris; Amaranthaceae) are rich sources of natural sugar.

Peanuts (Arachis) and soybeans (Glycine), both members of Fabaceae, the legume family, produce edible seeds that are important for their rich supply of protein or oil. Other plants rich in oil and important economically are the castor bean (Ricinus; Euphorbiaceae), coconut (Cocos nucifera; Arecaceae), flax (Linum usitatissimum; Linaceae), olives, oil palm (Elaeis guineensis; Arecaceae), sesame (Sesamum; Pedaliaceae), and sunflowers (Helianthus; Asteraceae).

Some plants produce toxic secondary compounds for protection. Some of the secondary compounds produced by angiosperms are not toxic, however; in fact, many are found in herbs and spices—for example, cloves, the dried flower buds of Syzygium aromaticum (Myrtaceae). The use of herbs and spices in cooking predates recorded history. Herbs are usually leaves or young shoots of nonwoody plants, although bay leaves and a few other leaves from woody plants are also considered herbs. Spices are the highly flavoured, aromatic parts of plants that are usually high in essential oil content. Spices are derived from roots, rhizomes, bark, seeds, fruits, and flower parts. The search for spices and alternative shipping routes for spices played a major role in world exploration in the 13th to 15th century. Many beverages are also derived from angiosperms; these include coffee (Coffea arabica; Rubiaceae), tea (Camellia sinensis; Theaceae), many soft drinks (e.g., root beer from the roots of Sassafras albidum; Lauraceae), and most alcoholic beverages (e.g., beer and whiskey from cereal grains and wine from grapes).

Coffee (Coffea arabica).

The angiosperms provide valuable pharmaceuticals. With the exception of antibiotics, almost all medicinals either are derived directly from compounds produced by angiosperms or, if synthesized, were originally discovered in angiosperms. This includes some vitamins (e.g., vitamin C, originally extracted from fruits); aspirin, originally from the bark of willows (Salix; Salicaceae); narcotics (e.g., opium and its derivatives from the opium poppy, Papaver somniferum; Papaveraceae); and quinine from Cinchona (Rubiaceae) bark. Some angiosperm compounds that are highly toxic to humans have proved to be effective in the treatment of certain forms of cancer, such as acute leukemia (vincristine from the Madagascar periwinkle, Catharanthus roseus; Apocynaceae), and of heart problems (digitalis from foxglove, Digitalis purpurea; Plantaginaceae). Muscle relaxants derived from curare (Strychnos toxifera; Loganiaceae) are used during open-heart surgery.

Foxglove (Digitalis).

The contribution of the angiosperms to biodiversity and habitat is so extremely important that human life is totally dependent on it. A significant loss of angiosperms would reduce the variety of food sources and oxygen supply in a habitat and drastically alter the amount and distribution of the world's precipitation. Many sources of food and medicine doubtless remain to be discovered in this group of vascular plants.

Structure and Function

Vegetative Structures

There are three levels of integrated organization in the vegetative plant body: organ, tissue system, and tissue. The organs of the plant—the roots, stems, and leaves—are composed of tissue systems (dermal tissue, ground tissue, and vascular tissue). The tissues of each of these systems are composed of cells of one or more types (parenchyma, collenchyma, and sclerenchyma). Tissues composed of only one cell type and performing only one function are simple tissues, while those composed of more than one cell type and performing more than one function, such as support and conduction, are complex tissues. Xylem and phloem are examples of complex tissues.

The plant develops from a fertilized egg, called a zygote, which undergoes mitotic cell division to form an embryo—a simple multicellular structure of undifferentiated cells (i.e., those that have not developed into cells of a specific type)—and eventually a mature plant. The embryo consists of a bipolar axis that bears one or two cotyledons, or seed leaves; in most dicots the cotyledons contain stored food in the form of proteins, lipids, and starch, or they are photosynthetic and produce these products, whereas in most monocots and some dicots the endosperm stores the food and the cotyledons absorb the digested food. The embryos of dicotyledons have two seed leaves, while those of monocotyledons have only one.

As the embryo continues to develop and new cells arise, the angiospermous plant develops specialized regions in which only cell division takes place and other areas in which nonreproductive (vegetative) activities, such as metabolism, respiration, and storage, occur. The areas of dividing cells, essentially permanently embryonic tissue, are called meristems, and their cells are termed initials. In the embryo they are found at either end of the bipolar axis and are called apical meristems. As the plant matures, apical meristems in the shoots produce new buds and leaves, and apical meristems in the roots are the points of active growth for roots. All growth produced by the apical meristems is primary growth and results in more primary tissues, which essentially extends the primary plant body.

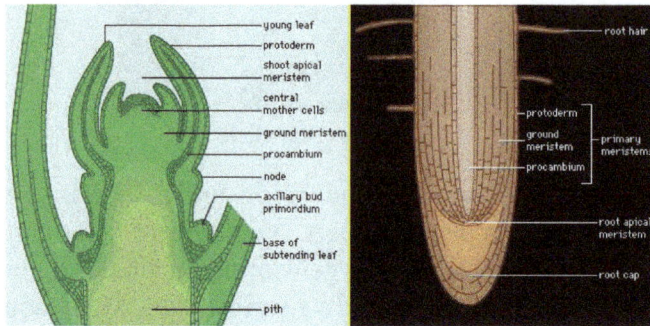

Apical meristems. (Left) The shoot apical meristem of Hypericum uralum appears at the topmost aspect of the stem. Immediately behind the apical meristem are three regions of primary meristematic tissues. (Right) The root apical meristem appears immediately behind the protective root cap. Three primary meristems are clearly visible just behind the apical meristem.

After a cell in an apical meristem has divided mitotically, one of the two resulting daughter cells remains in the meristem as an initial cell, and the other cell is displaced into the plant body as a derivative cell. The displaced derivative cell may divide several times as it differentiates (changes in structure and physiology) from a meristemic cell into a mature cell, but only initial cells remain permanently in the apical meristem. However, although most permanently differentiated derivative cells are nondividing cells, and regions of division remain in the root and shoot apical meristems, there are regions of dividing derivative cells behind apical meristems that give rise to primary tissue systems and thus are also considered to be primary meristems.

Three concentric regions of primary meristematic tissues develop immediately behind the apical meristem. These primary meristems produce the different tissues of the plant body: the outermost protoderm differentiates into the epidermis, a tissue that protects the plant; the adjacent ground meristem differentiates into the central ground tissues (the pith and cortex); and the procambium differentiates into the vascular tissues (the xylem, phloem, and vascular cambium). The xylem and phloem are conducting and supporting vascular tissues, and the vascular cambium is a lateral meristem that gives rise to the secondary vascular tissues, which constitute the secondary plant body.

Lateral meristems, called cambia, run the length of the stems and roots of vascular plants and produce secondary tissues, which develop after a plant organ—or part of a plant organ—has ceased to elongate. Secondary growth is essentially an increase in girth. The vascular cambium produces secondary xylem and secondary phloem, and the cork cambium (phellogen) produces cork cells, from which the outer bark develops. Figure summarizes the patterns of primary and secondary growth from root and shoot apical meristems.

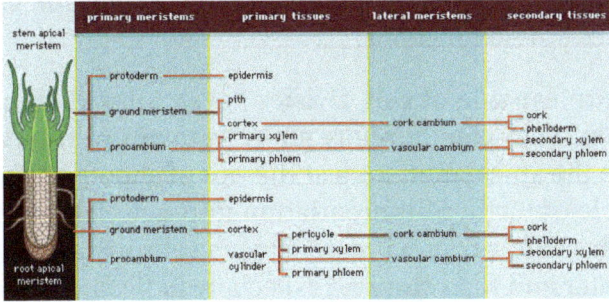

A summary of the primary and secondary growth of a woody dicotyledon.

Tissue Systems

three areas of meristematic tissue are derived directly from the apical meristem: the ground meristem, procambium, and protoderm. These meristematic tissues differentiate into the three primary tissues that constitute the primary plant body: ground tissue (pith and cortex), vascular tissue (xylem, phloem, and eventually the lateral, or secondary, meristem called the vascular cambium), and dermal tissue (epidermis), respectively.

Ground Tissue

The ground tissue system arises from a ground tissue meristem and consists of three simple tissues: parenchyma, collenchyma, and sclerenchyma. The cells of each simple tissue bear the same name as their respective tissue.

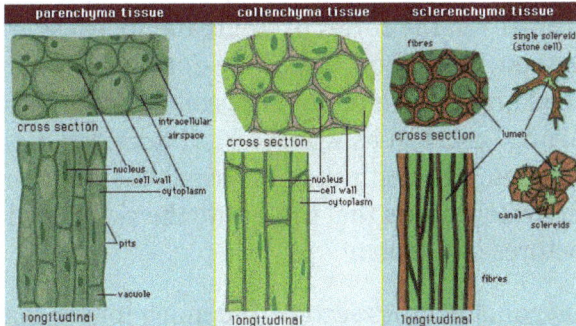

Cell types and tissues.

Parenchyma, often the most common ground tissue, takes its name from the Greek para, meaning beside, and egchnma, meaning the contents of a pitcher (literally, something poured beside), indicating its ubiquitous nature throughout the plant body. It forms, for example, the cortex and pith of stems, the photosynthetic tissue layer within the epidermis of the leaves (mesophyll), the cortex of roots, the pulp of fruits, and the endosperm of seeds. Parenchyma is composed of relatively simple, undifferentiated parenchyma cells. In most plants, metabolic activity (such as respiration, digestion, and photosynthesis) occurs in these cells because they, unlike many of the other types of cells in the plant body, retain their protoplasts (the cytoplasm, nucleus, and cell organelles) that carry out these functions.

Parenchyma cells are capable of cell division, even after they have differentiated into the mature form. They can therefore give rise to adventitious buds and roots at some distance from the apical meristem at the tips of shoots and roots. Parenchyma cells are also capable of further differentiation into new cell types under appropriate conditions, such as after trauma. Parenchyma cells are active in secretion, photosynthesis, and water and food storage (especially in fleshy fruits). They have large fluid-filled vacuoles that maintain cell turgidity; when a plant wilts, for example, it is because the vacuoles in the parenchyma cells have lost water and have become flaccid. Thus, parenchyma also functions in plant support. However, parenchyma cells do not have a secondary cell wall at maturity and thus remain flexible and capable of elongation.

Prosenchyma cells are starch-containing parenchymal cells whose cell walls have become lined with lignin, as occurs in the stems of Bougainvillea (Nyctaginaceae). A specialized type of parenchyma cell, called a transfer cell, is involved in the short-distance movement of solutes by cell-to-cell transfer. Transfer cells occur in association with veins in leaves and stems and also in many reproductive parts.

Collenchyma tissue consists of collenchyma cells that also have retained their protoplasts. They are closely related to parenchyma, although they have thick deposits of cellulose in their primary cell walls, and the two types often intergrade in areas of continuity.

Collenchyma is found chiefly in the cortex of stems and in leaves. For many herbaceous plants it is the chief supporting tissue, especially during early stages of development. In plants in which secondary growth occurs, the collenchyma tissue is only temporarily functional and becomes crushed as woody tissue develops. Collenchyma is located along the periphery of stems beneath the epidermal tissue. It may form a complete cylinder or occur as discrete strands that constitute the ridges and angles of stems and other supporting structures of the plant.

Collenchyma cells, polygonal in cross section, are much longer than parenchyma cells. The strength of the tissue results from the thickened cell walls and the longitudinal overlapping and interlocking of the cells. The wall is not uniformly thick in all cells, and thickening may occur predominantly in longitudinal strips at the corners of the cell, on the tangential (i.e., outer, toward the stem exterior) surface of the cell, or around the spaces between adjacent cells. Pits are present in the cell wall and provide a mechanism for intercellular communication. An important feature of collenchyma is that it is extremely plastic—the cells can extend and thus adjust to increase in growth of the organ. Because collenchyma cells are alive at maturity, these thickenings may be reduced when meristematic activity is resumed as in formation of a cork cambium or in response to wounding.

Sclerenchyma tissue is composed of sclerenchyma cells, which are usually dead at maturity (i.e., have lost their protoplasts). They characteristically contain very thick, hard secondary walls lined with lignin; consequently, sclerenchyma provides additional support and strength to the plant body.

The two principal types of sclerenchyma cells are sclereids and fibres. Sclereids vary in shape and size and may be branched. They are common in seed coats and nutshells. Apart from providing some internal support for various plant organs, sclereids deter desiccation of hard seeds, such as beans, and discourage herbivory of certain leaves.

Fibres are slender cells, many times longer than they are wide. They are highly lignified cells with tapering (oblique) end walls. The side walls of fibres are often so thick that the centre of the cell (the lumen) is often occluded. Fibres have great tensile strength and yet are also elastic. These qualities are significant in the flexible support of the stems of large herbs and leaves of many monocotyledons, such as palms. Leaf fibres are the source of abaca, or Manila hemp (Musa textilis; Musaceae), sisal (Agave sisalana; Asparagaceae), and many other fibre products. Fibres are found in various parts of the plant and are particularly common in the vascular tissues.

Vascular Tissue

Evolution of the Transport Process

Water and nutrients flow through conductive tissues (xylem and phloem) in plants just as the bloodstream distributes nutrients throughout the bodies of animals. This internal circulation, usually called transport, is present in all vascular plants, even the most primitive ones.

The importance of transport processes in plants increased as multicellular plants evolved and became larger and their tissues acquired specialized functions. As land plants developed, long-distance transport assumed an important role; not only are carbohydrates transported from the organs in which they are formed (the leaves) to other parts—such as reproductive organs (flowers and fruits), stems, and roots—but water and minerals must be transported to leaves, which are not submerged in water (as are those of most primitive nonvascular plants) but are in a relatively dry air environment. Highly developed land plants have two types of tissues specialized for long-distance transport: the xylem and the phloem. Water and dissolved mineral nutrients ascend in the xylem (the wood of a tree, such as an oak or a pine), and products of photosynthesis, mostly sugars, move from leaves to other plant parts in the phloem (the inner bark of a tree).

Evolving land plants faced not only the problem of transport but also the problem of supporting their weight. Aquatic plants are supported by their buoyancy in water and do not need a rigid stem; flotation devices such as gas-filled stomata and intercellular spaces hold them upright and enable them to grow toward the water surface and obtain sufficient sunlight for photosynthesis. On land, a rigid, self-supporting structure is necessary for plants; this structure, the xylem, consists of tiny rigid tubes through which water and dissolved mineral nutrients can move. The rigidity of the tubes within a stem is sufficient to make it self-supporting.

Land plants take up water from the soil through the roots; some exceptions, such as some desert plants that grow in dry soil and epiphytes, which grow in tree canopies, rely on adaptations that enable them to obtain water from the air. In most plants, then, water ascends through the xylem, the tiny capillaries of the woody stem tissue, into all plant parts but primarily into the leaves, from which it is transpired (evaporated) into the air. In this way, the mineral nutrients are transferred from the soil to all aboveground plant parts.

Tillandsia aeranthos.

Plants living in humid habitats, such as the small and primitive mosses and liverworts, do not have a well-developed xylem, but rather have similar cells called hydroids that lack true lignin. Similarly, water plants that have returned from land to an aquatic habitat during evolution have a reduced xylem; such plants, which have readapted to an aquatic environment, are not woody, because they need neither water-conducting tissues nor a self-supporting structure. On the other hand, tall land plants such as trees, vines, and lianas have the most highly developed long-distance transport systems. Vines and lianas differ from trees in that their xylem serves primarily for water conduction; they depend, for the most part, on other plants for support. Certain vines are of great length (a few hundred metres) and have extremely highly developed tissues for transporting water and nutrients.

Most of the material that composes a plant's dry weight is a consequence of photosynthesis, in which light energy is converted into chemical energy used to synthesize organic substances. Carbon dioxide from the air and water, which the plant takes from the soil, are utilized during photosynthesis, which occurs mostly in green plant parts—especially the leaves. Since plants shed their leaves either continuously or periodically but still increase in size, it is clear that many photosynthetic products must be transported out of the leaves and carried to all other plant parts; this process takes place primarily in the phloem.

The discovery of the functions of xylem and phloem was made following that of the circulation of blood in the 17th century. By the early 19th century, it had been established that water ascends from roots into leaves through xylem and that photosynthetic products descend through phloem. Experiments now called girdling experiments were performed, in which a ring of bark is removed from a woody plant. Girdling, or ringing, does not immediately interfere with upward movement of water in the xylem, but it does interrupt phloem movement. In some plants surgical removal of phloem is difficult; in this case phloem may be killed by using steam (steam girdling). Xylem conduction is normally not affected by such treatment, and movement in the two transport tissues can thus easily be distinguished. Girdling experiments, however, are not entirely foolproof. The question as to whether or not mineral nutrients can ascend in the phloem illustrates the kinds of difficulties that may be encountered. Much smaller amounts of mineral nutrients reach the leaves in girdled plants than in ungirdled ones. From this observation it might be concluded that some nutrients ascend in the phloem of ungirdled trees; girdling, however, interrupts the flow of sugars into roots. Roots are thereby starved and take up fewer mineral nutrients; the reduced flow of mineral nutrients to the leaves of girdled plants can thus be explained as a secondary effect.

Structural basis of Transport

Two features of plant cells differ conspicuously from those of animal cells. In plant cells the protoplast, or living material of the cell, contains one or more vacuoles, which are vesicles containing aqueous cell sap. Plant cells are also surrounded by a relatively tough but elastic wall. Water entering the vacuole by osmosis (i.e., movement of water across a membrane from regions of higher water concentration into regions of lower water concentration that normally contain dissolved substances, such as cell interiors) expands the protoplast and consequently the cell wall until the internal pressure is balanced by the elastic counterpressure of the wall. Spaces between and within cell walls are sufficiently large to permit water to flow around all cells. The space available for free water flow is called apoplast. Water in apoplast originates from the roots and contains nutrients taken up by them. Nutrients enter a cell by crossing the outer cytoplasmic membrane (plasma membrane).

Most of the metabolic activities of the cell—the chemical reactions of living systems—occur within protoplasts. Substances can enter a protoplast by their cytoplasmic connections between neighbouring cells (plasmodesmata) or by active transport mechanisms requiring energy and a group of enzymelike compounds called permeases.

Plasmodesmata may penetrate neighbouring cell walls at areas called primary pit fields. Also, some substances pass out of cells into the apoplast and are transported by energy-requiring processes into the protoplast of another cell.

Cell-to-cell transport takes place in all plants, but it is a slow process; the higher plants evolved the specialized tissues, xylem and phloem, for rapid long-distance transport. The woody tissue, xylem, contains highly specialized cells for water conduction. The cells are long and reinforced by strong, woody (lignified) walls; their protoplast breaks down and dissolves after wall growth is completed, so that the entire inside of the cell becomes available for rapid water conduction. In other words, the water-conducting cells of xylem are dead when functional. In the more primitive conifers the xylem consists largely of spindle-shaped cells called tracheids, which have a diameter around 0.04 millimetre (0.0016 inch) and a length of about 3 millimetres (0.12 inch). Flowering plants have a more highly specialized xylem, in which the mechanical function and the water-conduction function have been separated during evolution. Tracheids, the primitive conducting cells, have evolved into fibres for mechanical strength and vessels for water conduction, particularly in angiosperms. Vessel elements are barrellike cells with widths of up to 0.5 millimetre (0.02 inch) in some plants. Vessel elements are arranged end to end; their end walls are partly or wholly dissolved, and rows of such cells thus form long capillaries (tubes) up to several metres in length. These tubes are the vessels.

Eucalyptus tree.

Numerous vessels of limited length thus provide a certain protection against injury—that is, since water pressures in the xylem are often well below zero (i.e., the water is under tension), air will be sucked into any injured xylem vessel and spread immediately throughout it but cannot pass through the wet pit membranes into the uninjured units. Damage is thus confined to the units that are injured and cannot easily spread. In addition, the smaller the conducting unit, the more confined is the damage. Plants with large, highly efficient vessels are much more vulnerable to injury, as is evident, for example, from the vulnerability of the elm, which has large vessels, to Dutch elm disease, in which the water-conduction vessels are injured by beetle activity and fungal growth. In general, both the less efficient but safer coniferous wood and the more highly efficient but more vulnerable wood of flowering plants have been successful during evolution. Very tall trees occur in both groups—e.g., Sequoia among the conifers and Eucalyptus among the flowering plants.

The conducting elements of the phloem underwent evolutionary changes somewhat similar to those of the xylem. The conducting elements of conifers, called sieve cells, are similar in shape and dimensions to tracheids. They do not have a woody wall, however, and they are alive at functional maturity even though their cytoplasm may be highly specialized and the cells have usually lost their nucleus during development. In flowering plants the conducting elements in the phloem are called sieve elements and consist of sieve cells and sieve-tube members, the latter differing in having some sieve areas specialized into sieve plates (generally on the end walls). Sieve-tube members are arranged end to end to form sieve tubes, a name derived from the sievelike end walls through which passage of food from one cell to the next occurs. Sieve elements are almost invariably accompanied by special companion cells believed to control, to a certain extent, the metabolism of the nucleus-free conducting cells.

Organization of the Vascular Tissue

Vascular tissue is organized into discrete strands called vascular bundles, each containing xylem and phloem. In stems, the vascular tissue is organized into many discrete vascular bundles. In the roots, the vascular tissue is organized within a single central vascular cylinder.

The xylem conducts water and minerals within the primary plant body, and the phloem conducts food. The xylem cells are arranged end to end to form a longitudinal continuum throughout the plant. The phloem cells form a similar continuum. Thus, water enters the xylem cells in the roots and travels to the leaves via the stems, and photosynthates (products of photosynthesis) enter the phloem cells in the leaves and are translocated to the roots via the stems. Storage parenchyma and fibres are generally present, and sclereids rarely are.

Primary xylem consists of lignified tracheary elements (tracheids and vessel elements), which are dead at maturity (they have lost their protoplasts). Parenchyma cells also are interspersed throughout the tissue. Both tracheids and vessel elements are long hollow cells with tapered end walls. The end walls of adjacent tracheids contain paired small, rimmed, nonperforated pores, called bordered pits; water diffuses through a shared central membrane. The side walls have five patterns of thickening, which are believed to represent a developmental sequence from the initial xylem (protoxylem) to the final mature xylem (metaxylem): annular (a series of rings), helical (a long continuous spiral), reticular (a network), scalariform (a series of elongated bordered slits), and circular bordered pitting. Individual species may omit some of these patterns.

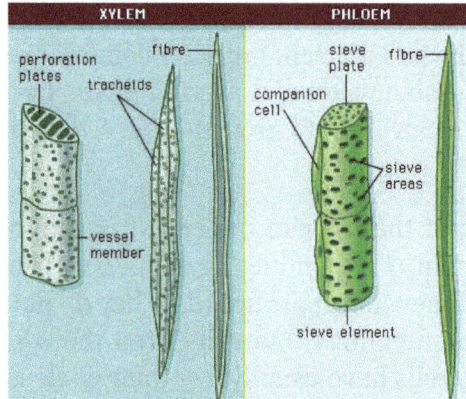

Cells of the xylem and phloem.

Vessel elements differ from tracheids in that the end walls are modified into perforation plates, an area or areas in which there is no shared wall material or membrane. Vessel elements join to form continuous vessels. The perforations are much larger than those of the bordered pits of tracheids and are of four types: scalariform (slitlike), foraminate (circular), reticulate (a network), or simple (single). The bordered pitting of the side walls of vessel members is either scalariform or circular (generally scalariform bordered pitting is associated with scalariform, foraminate, or reticulate perforation plates). Vessel elements are found in the late metaxylem (the final, or most developed, form of the primary xylem).

The most common type of perforation plates in the angiosperms are scalariform and simple; the other types are rare. The putatively primitive angiosperms are without vessels and evolved from a condition in which only tracheids were present to one in which a series of long vessel elements had scalariform lateral walls and highly inclined end walls with many scalariform perforations, to short vessel elements with circular bordered pits in lateral walls and simple perforation plates in horizontal end walls.

This series of specializations has increased the efficiency with which water moves through the vessels: from the more generalized method of water diffusion through pit membranes of narrow tracheids to mass movement of water through the perforated end walls of relatively narrow scalariform vessels and then to relatively wide simple vessels with large single perforated end walls. This simple form is a rather streamlined system that facilitates the maximum movement of water in terms of amount and speed with the minimum amount of resistance, allowing for greater efficiency and effective water transport.

The primary phloem is composed of sieve elements and fibres. Parenchyma cells are interspersed throughout. Sieve elements are longitudinal cells that transport food. They are composed of sieve cells and sieve-tube members. Sieve-tube members have clusters of pores in the cell walls known as sieve areas, which have either small pores or large pores; the latter are known as sieve plates. Sieve plates are mostly located on the overlapping adjacent end walls. As sieve-tube members differentiate, they lose their nucleus, ribosomes, vacuoles, and dictyosomes (the equivalent of the Golgi apparatus in animals); they are not dead, however, and remain metabolically active. Each sieve-tube member has an associated specialized parenchyma cell called a companion cell. They are derived by mitosis from the same parent cell and remain connected with each other. Photosynthates are actively secreted into, and actively removed from, sieve-tube members by their companion cells. Other unspecialized parenchyma cells also are present in primary phloem and provide storage.

Finally, the primary vascular tissue system usually has fibres, particularly in herbaceous plants. The fibres occur in groups either around vascular bundles or as a cap over the phloem (phloem fibres).

The primary vascular system serves three functions. First, the sieve tubes conduct photosynthates via companion cells from green stems and leaves to nongreen areas (usually roots, lateral meristems, and shoot apical meristems) to promote growth and development. Second, tracheary elements provide a water-conducting system and a support system as a result of their rigid lignified cell walls. Third, fibres provide additional support.

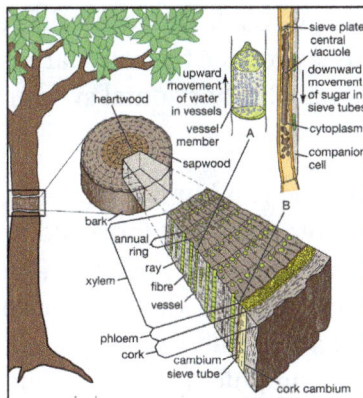

Internal transport system in a tree. (A) Enlarged xylem vessel.
(B) Enlarged mature sieve element.

Secondary Vascular System

In woody plants, a vascular system of secondary vascular tissue develops from a lateral meristem called the vascular cambium. The vascular cambium, which produces xylem and phloem cells, originates from procambium that has not completely differentiated during the formation of primary xylem and primary phloem. The cambium is thought to be a single row of cells arranged as a cylinder that produces new cells: externally the secondary phloem and internally the secondary xylem. Because it is not possible to distinguish the cambium from its immediate cellular derivatives, which also divide and contribute to the formation of secondary tissues, the cambium and its immediate derivatives are usually referred to as the cambial zone.

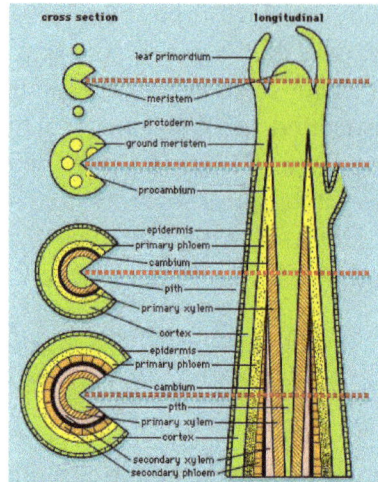

Tissue organization in a stem tip.

Unlike the apical meristems, which consist of a population of similar cells, the cambium consists of two different cell types; the fusiform initials and the ray initials. The fusiform initials are elongated tapering cells that give rise to all cells of the vertical system of the secondary phloem and xylem (secondary tracheary elements, fibres, and sieve cells and the associated companion cells). The ray initials are isodiametric cells—about equal in all dimensions—and they produce the vascular rays, which constitute the horizontal system of secondary tissues; this horizontal system acts in the translocation and storage of food and water.

The fusiform and ray initials of the cambium divide in a plane tangential to the surface of the stem, with the long axes of the fusiform and ray initials parallel to the long axis of the plant organ. The cambium generates xylem mother cells toward the inside and phloem mother cells toward the outside. These cells in turn continue to divide tangentially, producing new cells that add to the xylem and to the phloem. Divisions of the cambium cells and xylem and phloem mother cells do not result in the production of equal amounts of secondary xylem and secondary phloem; because the cambium produces more cells internally than externally, more secondary xylem is produced than secondary phloem. Because divisions in the fusiform and ray initials are primarily tangential, new cells are regularly arranged in well-defined radial rows, a characteristic pattern for secondary vascular tissues.

Divisions in the cambium not only produce secondary vascular tissues but also increase the circumference of the cambium. As new cells are continuously added to the inside of the cambium, the cambium increases laterally (in circumference) to keep pace with the circumferential growth of the stem. In some plants, this is accomplished simply by radial division of the fusiform and ray initials. In other plants, the mechanism for increasing cambial diameter or increasing the number of cambial cells is more complex. If cambial activity is extensive, the primary tissues lying outside the cambium, such as primary phloem, cortex, and epidermis, are crushed by the pressure of new secondary tissue growth or become torn and obliterated because they cannot accommodate the rapidly increasing diameter of the plant.

As growth proceeds, the cork cambium forms in living cells of the epidermis, cortex, or, in some plants, phloem and produces a secondary protective tissue, the periderm. The cork cambium is, like the vascular cambium, a lateral meristem that produces cells internally and externally by tangential divisions. Unlike the cambium, the cork cambium consists of one cell type.

Another type of meristem active in certain plants, especially grasses, is the intercalary meristem. These cells possess the ability to divide and produce new cells, as do apical and lateral meristems. They differ, however, in being situated between regions of mature tissue, such as at the base of grass leaves, which are themselves located on mature stem tissue. In many instances intercalary meristems function for only a short time and eventually completely differentiate into mature tissues. Intercalary meristems are usually located at positions on the stem where leaves have emerged (nodes) and are largely responsible for elongation in grass shoots and leaves. Intercalary meristems are the internode regions where cell division of the ground meristem persists for a longer time than in other areas of the internode. In rosette plants, intercalary meristems are lacking.

Secondary xylem is composed of tracheary elements, rays, fibres, and interspersed axial parenchyma cells. The tracheary elements consist of only tracheids, as in the few vessel-less angiosperms (e.g., Winteraceae), or of both tracheids and vessel elements, as in the vast majority of angiosperms. Axial parenchyma may surround the vessel elements (paratracheal) or be randomly dispersed among the vessel elements (apotracheal).

Tyloses are balloonlike outgrowths of parenchyma cells that bulge through the circular bordered pits of vessel members and block water movement. The presence of tyloses in white oaks makes their wood watertight, which is why it is preferred in casks and shipbuilding to red oak, which lacks tyloses and does not hold water. In trunks and branches that lean, there is eccentric growth of tension wood on the upper surface; tension wood is a type of reaction wood found in angiosperms that contains gelatinous fibres which shrink and pull.

Growth rings in the secondary xylem of temperate woody angiosperms are usually annual, but under environmental fluctuations, such as drought, more than one can form, or none at all. Growth rings result from the difference in density between the early wood (spring wood) and the late wood (summer wood); early wood is less dense because the cells are larger and their walls are thinner. Although the transition of early wood to late wood within a growth ring may be obscure, that demarcation between the adjacent late wood of one ring and the early wood of the next ring is clear. Diffuse-porous wood occurs when the size of the vessels (pores) in a growth ring are fairly uniform and evenly distributed (e.g., red maple, Acer rubrum; Sapindaceae). Ring-porous wood occurs when the pores of the early wood are distinctly larger than those of the late wood (e.g., black walnut, Juglans nigra; Juglandaceae).

Both xylem and phloem have limited longevity. The oldest phloem layers are the outermost—the dead bark of the stem surface. The yearly amounts of xylem visible as distinct rings in cross sections of stems are known as annual rings. The oldest xylem layers (i.e., the oldest annual rings) are in the dead central core, or heartwood, of the woody stem, which can often be recognized by its darker coloration. The lighter-coloured sapwood is living and functions as storage tissue and, especially in the outermost sapwood, as conducting tissue; the younger annual rings make up the sapwood. In some highly specialized tree species with large vessels (such as some oaks, ashes, and others), only the very outermost growth ring functions in water conduction.

Conducting tissues seldom run straight along a tree stem; usually they are arranged in a helical or spiral pattern, sometimes called the spiral grain of a tree. The angle of the spiral arrangement usually changes from year to year; the path of water up a tree stem may therefore be very complicated if more than one growth layer acts as a conducting tissue. Functionally, the effect of the variable spiral grain is to distribute water to all parts of the tree from any root.

The secondary phloem of angiosperms consists of sieve-tube members, companion cells, scattered parenchyma, ray parenchyma, and fibres. The fibres usually occur in clusters or as bands alternating with bands of sieve tubes and parenchyma cells. As the vascular cambium continues to produce more secondary xylem to the inside, the older (most exterior) portions of the secondary phloem are crushed, die, and are sloughed off as part of the bark. Successive cork cambiums, essentially lateral meristems from which the bark arises, originate in the parenchyma of the phloem and produce additional cork.

Uptake of Water and Mineral Nutrients from the Soil

Water uptake from the soil by root cells is passive, in that water may be pulled into the root by low xylem pressure and also follows osmotic gradients caused by the mineral nutrients, which are taken up actively (i.e., with the expenditure of metabolic energy) across root cell membranes. As the mineral nutrients—the ions (charged components) of inorganic salts—are taken up, they are largely incorporated into organic molecules. Thus, the solutes in xylem sap are mostly complex organic substances, sometimes of a specific nature; for example, nicotine synthesis takes place in the roots of tobacco plants, where nitrogen is incorporated into compounds that have moved to the roots through the phloem as sugars. If a tomato shoot is grafted onto a tobacco rootstock, nicotine-containing tomato leaves are formed. On the other hand, a tobacco shoot grafted onto a tomato rootstock results in a plant with nicotine-free tobacco leaves. Many other specific nitrogen-containing substances originate in the roots; in most plants, however, nitrogen is transported to the leaves from the roots in the form of compounds known as amino acids and amides.

The major chemical elements needed by a plant are carbon, hydrogen, oxygen, phosphorus, potassium, nitrogen, calcium, iron, and magnesium; in addition, many other elements are required in very small amounts. A lack of any element may result in deficiency diseases. A few elements taken up by plants are of no nutritive value and usually are eliminated or crystallized (e.g., silica), sometimes by deposition in special cells.

The plant is able to control to some extent the substances that enter. If equal amounts of sodium and potassium are available to roots of plants, and the amount of the two elements inside the plant is analyzed, less sodium is likely to be found than potassium. The structural basis for the control of uptake of substances into roots is the so-called Casparian strip, a conspicuously thickened wall area one cell layer deep surrounding primary roots; it prevents excess soil solution from being pulled directly into the central part of the root where the xylem is located. As a result, the soil solution has to pass through a cell barrier in which uptake can be metabolically controlled. After nutrients are inside living root cells and have been converted to appropriate compounds, the latter are released into the xylem and move to above-ground parts.

Process of Xylem Transport

The total amount of conducting tissue remains about the same from roots to leaves. In terms of water movement, the velocity of movement might be expected to be uniform throughout the entire axial system of stem, branches, and twigs. Because some trees (e.g., oaks) have thick twigs, however, the velocity of water movement is greater in the stem than in the twigs at any time. Similarly, in tree species with slender branches (such as birches), the reverse is true. Normally the proportion of xylem to leaves supplied by that xylem is greater in plants growing in dry habitats than in plants found in wet ones and may be as much as 700 times greater in certain desert plants than in aquatic plants and herbs of relatively humid forest floors. The leaves of dry-habitat plants thus are more richly supplied with water-conducting xylem tissue than are those of moist habitats.

The velocity of sap movement in trees varies throughout a 24-hour period. During the night, especially a rainy night, sap flow may stop; velocity increases with daylight, peak rates being found in the early afternoon. Peak velocities correlate with vessel size; the rate of sap flow in trees with small vessels is about 2 metres (7 feet) per hour; that in trees with large vessels, about 50 metres (160 feet) per hour. The energy required to lift water in both cases is comparable; in trees with large pores, water simply moves faster through fewer and larger vessels.

It was demonstrated about 1900 that living cells of the stem are not responsible for water movement. It is now generally recognized that water in the xylem moves passively along a gradient of decreasing pressures. Under certain special conditions, water is pushed up the stem by root pressure. This may be the case with herbaceous (non-woody) plants in the greenhouse under conditions of ample water supply and little transpiration. In nature, these conditions may be met in early spring before the leaves emerge, when the soil is wet and transpiration is low. Under such conditions, water movement is caused by active uptake of ions (charged particles) and by the entry of water from the soil into the roots. Most of the time, however, water is pulled into the leaves by transpiration. A gradient of decreasing pressures from the base to the top of a tree can be measured, even though pressures are low.

A vacuum pump cannot pull water to a height of more than 10 metres (about 33 feet). Since many trees are far taller than 10 metres, the mechanism by which they move water to their crowns has been investigated. Is it possible for trees to pull water into their crowns along a decreasing pressure gradient or do they employ some other mechanism? If trees pull water, that in the xylem would have to be held on the tracheid and vessel walls by adhesion, and water molecules would have to hold together by cohesion. The hypothesis that water is pulled upward along a pressure gradient during transpiration has been called the cohesion theory. Two critical requirements of the cohesion mechanism of water ascent are (1) sufficient cohesive strength of water and (2) existence of tensions (i.e., pressures below zero) and tension gradients in stems of transpiring trees.

Although the tensile strength of water is very high, an excessive pull exerted on a water column will break it. The tallest trees are about 100 metres (330 feet) high. A nonmoving water column at an atmospheric pressure of 1 atmosphere at the base of the tree is exposed to a pressure of –9 atmospheres (i.e., a tension of 9 atmospheres) at the top. Under conditions of peak flow at midday, this gradient increases by about 50 percent; in other words, a transpiring sequoia would have a pressure in the xylem of at least –14 atmospheres at the top if the basal pressure is 1 atmosphere. If the pressure at the base is –10 atmospheres because of dry soil, however, the pressure at the top drops to –25 atmospheres. It has been demonstrated that water columns in the xylem can withstand this tension, or pull, without breaking.

Negative pressures and gradients of negative pressures have been shown to exist in trees with an ingeniously simple device called the pressure bomb. A small twig is inserted in a container (the pressure bomb), its cut stump emerging from a tightly sealed hole. As pressure is applied to the container and gradually increased, water from the xylem emerges from the cut end as soon as the pressure being applied is equal to the xylem tension that existed when the twig was cut. This method has been used to measure gradients of negative pressures in trees. Movement in the xylem is by mass flow of the whole solution, and the force is either the tension pull of transpiration or root pressure, or both. In general, however, water movement in the xylem is by transpiration pull. The mechanism of phloem transport remains unclear.

Process of Phloem Transport

Products of photosynthesis (primarily sugars) move through phloem from leaves to growing tissues and storage organs. The areas of growth may be newly formed leaves above the photosynthesizing leaves, growing fruits, or pollinated flowers. Storage organs are found in roots, bulbs, tubers, and stems. Thus the movement in the phloem is variable and under metabolic control (whereas movement in xylem is always upward from the roots).

The rate at which these substances are transported in the phloem can be measured in various ways—e.g., as velocities in distance traveled per unit time or as mass transfer in (dry) weight transported per unit time. Velocities appear to be graded—i.e., some molecules move faster than others within the same channel. Peak velocities of molecules usually are of the order of 100 to 300 centimetres (40 to 120 inches) per hour. Average velocities, more difficult to measure but significant in mass-transfer considerations, are lower.

Mass transfer can be measured by weighing a storage organ, such as a potato tuber or a fruit, at given time intervals during its growth. Mass transfer per cross-sectional area of conducting tissue is referred to as specific mass transfer and is expressed as grams per hour per square centimetre of phloem or sieve tubes. With a given specific mass transfer, the velocity with which a liquid of a certain concentration flows can be calculated; in dicotyledonous stems, for example, specific mass transfer is between 10 and 25 grams per hour per square centimetre of sieve tube tissue at times of peak performance. In certain tree species the sieve tubes can be tapped to obtain an exudate. The concentration of this exudate, multiplied by the measured average velocity, is of the same order of magnitude as specific mass transfer, indicating that liquid movement through sieve tubes could account for transport.

Much of the experimental work on phloem transport now is done with the aid of radioactive substances; for example, when radioactive carbon dioxide administered to an illuminated leaf is incorporated into sugar during photosynthesis and carried from the leaf, the velocity of this movement can be measured by determining the arrival of radioactivity at given points along the stem. Whole plants, as long as they are reasonably small, can be pressed against photographic film after the conclusion of a similar experiment, and the photographic image will indicate the areas to which radioactive sugar has moved.

The mechanism of phloem transport has been studied for many years. A number of hypotheses have been put forth over the past years, but none is entirely satisfactory. One fundamental question is whether sugars and other solutes move en masse as a flowing solution or whether the solvents diffuse independently of the solvent water. The phenomenon of exudation from injured sieve tubes supports the first possibility, which has been further supported by a discovery involving aphids (phloem-feeding insects): when aphids are removed from plants while feeding, their mouthparts remain embedded in the phloem. Exudate continues to flow through the mouthparts; the magnitude of the rate of this exudation indicates that transport within the sieve tube to the mouthparts occurs as a flow of solution.

Evidence against solution flow is the movement of substances in opposite directions through a section of phloem at any one given time. This, however, has never been convincingly demonstrated in just one sieve tube. On the other hand, attempts to find simultaneous movement of sugars and water along a phloem path, in order to demonstrate solution flow, have been only partially successful.

Mass-flow hypotheses include the pressure-flow hypothesis, which states that flow into sieve tubes at source regions (places of photosynthesis or mobilization and exportation of storage products) raises the osmotic pressure in the sieve tube; removal of sugars from sieve tubes in sink regions—i.e., those in which sugars are removed or imported for growth and storage—lowers it. Thus a pressure gradient from the area of photosynthesis (source) to the region of growth or storage (sink) is established in sieve tubes that would allow solution flow. The electroosmotic hypothesis postulates that solution is moved across all sieve plates (areas at which individual sieve elements end) by an electric potential that is maintained by a circulation of cations (positively charged chemical ions), such as potassium. Transport hypotheses postulating solute movement independent of solvent water include the spreading of solute molecules between two liquid phases and the active transport of molecules by a type of cytoplasmic movement that is often referred to as cytoplasmic streaming.

During the life of a leaf, its role as a sink or a source changes. A young developing leaf before it is photosynthetic is a sink for sugars produced by older leaves. After the leaf begins to expand and turn green, it is both a sink (importer of sugar) and a source (exporter of sugar) as a result of its own photosynthetic capacity. When mature and fully expanded, the leaf then becomes a source of sugar production.

Transport and Plant Growth

It is important to realize that the plant, with its two transport systems, xylem and phloem, is able to move any substance to virtually any part of its body; the direction of transport is usually opposite in the two systems, and transfer from one system to the other takes place easily. An exception is transport into flowers and certain fruits, in which flow in each system is unidirectional.

Numerous substances move from roots to mature leaves through xylem and are transferred from the leaves, together with sugars, through the phloem to other plant parts. In the autumn months in temperate regions, plants store most of the products resulting from photosynthesis during the summer months in structures such as stems, bulbs, and tubers and mobilize it in the spring when new growth begins. A few plants, such as some tropical monocotyledons (certain palms, for example), store food for many years for use at the time of flowering and fruit-set at the end of their lives.

Plant hormones, or growth regulators, are effective in very small amounts; they induce or enhance specific growth phenomena. Because the site of hormone synthesis is different from its place of action, hormones must be transported before they can exert their effects. There are five major types of plant hormones, including auxin, gibberellin, cytokinin, ethylene, and abscisic acid. Each type plays a different role in plant growth and development, from influencing cell division, fruit ripening, and seed dormancy to directing stem elongation and food mobilization. The best-characterized of these hormones are the auxins, the most common of which is called indoleacetic acid. Auxins are formed in young, growing organs, such as opening buds, and are transported away from tips of shoots toward the base of the plant, where they stimulate the cells to elongate and sometimes to divide. Responses to gravity and light are also under auxin control. Auxins move to the lower side of a leaning stem; cells on the lower side then elongate and cause the stem to bend back to a vertical position. Response to gravity in many roots is the opposite of that in shoots; the same mechanism of auxin distribution is responsible, but roots react to different quantities of the hormone than do shoots. Similar auxin distributions are responsible for phototropic responses—i.e., the growth of plant parts such as shoot tips and leaves toward light. In certain cases auxin may be destroyed on the illuminated side, and the unilluminated side with more auxin elongates, causing the shoot to bend toward the light.

Auxins are not normally transported through vascular tissue; moreover, transport is polar—i.e., it takes place along the stem from tip to base, regardless of the stem's position. Velocities of transport are of the order of 5 to 10 millimetres (0.2 to 0.4 inch) per hour, and transport requires the expenditure of metabolic energy. There is evidence that most growth hormones can be transported through xylem or phloem, but, at least in the case of auxin, the transport mechanism is specific directionally from morphological top to bottom.

Hormone transport is also involved in the stimulation of flowering. In some plants, flowering is triggered by short or long days. The receptor of this stimulus is in the leaves. A chemical substance, probably a flowering hormone of an as-yet-unknown nature, then moves to the shoot apex and causes a transformation of the vegetative growing point into a flowering shoot.

Many growth-correlating phenomena are effected by transported hormonal stimuli. A vigorously growing terminal (topmost) shoot may inhibit lateral buds lower down from growing out and may force later branches to bend down. If the terminal shoot is removed, laterals grow out and topmost lateral branches bend upward. In leaning trees with secondary tissue (wood), the cambium produces compression wood on the lower side (in conifers) or tension wood on the upper side (in dicotyledons) in response to a hormone; the stem responds by pushing (in conifers) or pulling (in dicotyledons) itself upright. Transport of growth-regulating substances is thus largely responsible for the characteristic shape of each plant species.

Dermal Tissue

The dermal tissue system—the epidermis—is the outer protective layer of the primary plant body (the roots, stems, leaves, flowers, fruits, and seeds). The epidermis is usually one cell layer thick, and its cells lack chloroplasts.

As an adaptation to a terrestrial habitat, the epidermis has evolved certain features that regulate the loss of water, carbon dioxide, and oxygen. Cutin and waxes are fatty substances deposited in the walls of epidermal cells, forming a waterproof outer layer called the cuticle. Often, epicuticular waxes, in the form of sheets, rods, or filaments, are exuded over the cuticle, giving some leaves their whitish, greenish, or bluish "bloom." The cuticle and epicuticular waxes minimize transpiration from the plant. The waxy deposits can be thin or thick, depending on the requirements of the plant; for example, desert plants usually have heavy wax coatings.

The plant, however, must have some means of exchanging water vapour, carbon dioxide, and oxygen through this cuticle barrier. Dispersed throughout the epidermis are paired, chloroplast-containing guard cells, and between each pair is formed a small opening, or pore, called a stoma (plural: stomata). When the two guard cells are turgid (swollen with water), the stoma is open, and, when the two guard cells are flaccid, it is closed. This controls the movement of gases, including water vapour in transpiration, into the atmosphere. Guard cells and stomata are found on aerial plant parts, most frequently on leaves, but are not known to occur on aerial roots.

Floral trichome Floral trichomes (plant hairs) on the buds and
sepals of thyme flowers (Thymus vulgaris).

The trichomes (pubescences) that often cover the plant body are the result of divisions of epidermal cells. Trichomes may be either unicellular or multicellular and are either glandular, consisting of a stalk terminating in a glandular head, or nonglandular, consisting of elongated tapering structures. Leaf and stem trichomes increase the reflection of solar radiation, thereby reducing internal temperatures, and thus reduce water loss in plants growing under arid conditions.

Epiphytic bromeliads (air plants such as Spanish moss, Tillandsia usneoides; Bromeliaceae) absorb water and minerals via foliar trichomes. The glandular trichomes produce and secrete substances such as oils, mucilages, resins, and, in the case of carnivorous plants, digestive juices. Plants growing in soils with high salt content produce salt-secreting trichomes (e.g., saltbush, Atriplex vesicaria; Amaranthaceae) that prevent a toxic internal accumulation of salt. In other cases, trichomes help prevent predation by insects, and many plants produce secretory (glandular) or stinging hairs (e.g., stinging nettle, Urtica dioica; Urticaceae) for chemical defense against herbivores. In insectivorous plants, trichomes have a part in trapping and digesting insects. Prickles, such as those found in roses, are an outgrowth of the epidermis and are an effective deterrent against herbivores.

The epidermis is the outermost protective layer of the primary plant body. At a certain stage in their life cycle, woody plants cease to grow in length and begin to add to their girth, or width. This is accomplished not by the addition of more primary tissue but by the growth of secondary vascular tissue around the entire circumference of the primary plant body. The secondary vascular tissue arises from the vascular cambium, a layer of meristematic tissue insinuated between the primary xylem and primary phloem. Secondary xylem develops on the inner side of the vascular cambium, and secondary phloem develops on the outermost side. A second lateral cambium, called the phellogen or cork cambium, is the source of the periderm, a protective tissue that replaces the epidermis when the secondary growth displaces, and ultimately destroys, the epidermis of the primary plant body.

Cork oak (Quercus suber) with sections of cork removed.

In woody plants, the phellogen, or cork cambium, arises in any of the three tissue systems near the surface of the plant body. The cork cambium produces cork cells toward the outside and parenchyma cells toward the inside. As a unit, the cork cambium, cork cells, and parenchyma (phelloderm) form the periderm. Like the epidermis, the periderm is a protective tissue on the periphery of the plant body; however, because the periderm is produced by a lateral meristem, it is considered to be of secondary origin (in contrast to the primary origin of the epidermis from the protoderm). At maturity the cork cells are nonliving, and their inner walls are lined with suberin, a fatty substance that is highly impermeable to gases and water (which is why cork is used to stop wine bottles). The walls of cork cells may also contain lignin.

In stems, the first cork cambium usually arises immediately inside the epidermis or in the epidermis itself. In roots, the first cork cambium appears in the outermost layer of the vascular tissue system, called the pericycle .

The meristematic tissue of the cork cambium produces more and more derivatives of cork cells and parenchyma and displaces them into the outer margins of the plant body. Because the epidermal cells do not divide, they cannot accommodate an increase in stem diameter. Thus, the epidermal cells soon become crushed by the growing number of cork cells derived from the cork cambium, eventually die, and are sloughed off.

The epidermis is then replaced by cork cells until eventually the original cork cambium ceases to produce derivative cork and is itself destroyed. A new cork cambium eventually arises in the secondary phloem situated just behind the old cork cambium. That portion of the secondary phloem that forms between the new cork cambium and the old one becomes crushed and displaced externally as well. This process is repeated often each growing season.

The term cork is used to denote the outer derivatives of the cork cambium specifically. Bark, on the other hand, is an inclusive term for all tissues outside of the vascular cambium. The two regions of the bark are the outer bark, composed of dead tissues, and the inner bark, composed of living tissues of the secondary phloem. Outer bark is shed continually from a tree, often in a distinctive pattern, as the circumference increases because its dead cells cannot accommodate the increased diameter. Bark contributes to the support of the tree and protects the living tissue of the active secondary phloem and vascular cambium from desiccation and from such environmental disturbances as fire.

Plant organs

Roots

The root apical meristem, or root apex, is a small region at the tip of a root in which all cells are capable of repeated division and from which all primary root tissues are derived. The root apex is protected as it passes through the soil by an outer region of living parenchyma cells called the root cap. As the cells of the root cap are destroyed and sloughed off, new parenchyma cells are added by a special internal layer of meristematic cells called the calyptrogen. Root hairs also begin to develop as simple extensions of protodermal cells near the root apex. They greatly increase the surface area of the root and facilitate the absorption of water and minerals from the soil.

Along the longitudinal axis of a root, beginning with the root cap and leading away from the root tip, there are five distinct zones in which certain specific growth patterns dominate: cell division, cell elongation, primary tissue maturation, mature primary tissues, and secondary tissue growth (the latter is found in woody roots—i.e., those of perennial dicotyledons). There is a gradual transition between these regions.

The region of cell division includes the apical meristem and the primary meristems—the protoderm, ground meristem, and procambium—derived from the apical meristem. As is generally true of nonmeristematic regions elsewhere in the plant body, root length in the second region is increased as a result of cell elongation rather than by cell division. The region of maturation that follows is where the cells differentiate (i.e., change in structure and physiology into cells of a specific type) and where the first primary phloem and xylem, as well as mature root hairs, are clearly seen. The region of mature primary tissues is where the anatomy of the primary body of the root is most obvious and where all the elements of the vascular cylinder, cortex, and epidermis are evident. Finally, in the region of secondary growth, the secondary xylem and phloem as well as the periderm add girth to the plant.

There are many individual vascular strands (or vascular bundles) in the primary body of the stem , and they all converge into a single central vascular cylinder in the root, forming a continuous system of vascular tissue from the root tips to the leaves. At the centre of the vascular cylinder of most roots is a solid, fluted (or ridged) core of primary xylem. The primary phloem lies between these flutes or ridges. Parenchyma cells are dispersed throughout the vascular cylinder.

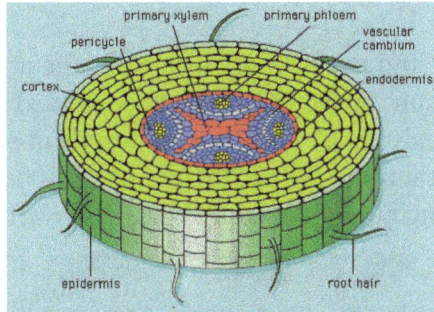

Cross section of a typical root, showing the primary xylem
and phloem arranged in a central cylinder.

The vascular cylinder of the root is surrounded by a layer of parenchymatous pericycle cells. As the root ages, many of these cells become fibres, particularly in monocotyledons and many herbaceous dicotyledons. As defined above, a characteristic feature of parenchyma cells is their ability to differentiate into cells of a different type under appropriate conditions. The parenchyma cells of the pericycle, then, can be considered meristematic in that they give rise to new lateral meristems and lateral roots. In woody roots the vascular cambium (the lateral meristem that gives rise to secondary phloem and secondary xylem) originates in the pericycle as well as in the procambium; the procambium is the primary meristematic tissue between the primary phloem and xylem. The first cork cambium is a lateral meristem that arises in the pericycle; the successive cork cambia arise in the secondary phloem. Because lateral roots are initiated in the pericycle and grow out through the cortex and epidermis, they are said to have an internal, or endogenous, origin, in contrast to the external, or exogenous, origin of leaves and the apical meristem of stems.

Ground tissue called the cortex surrounds the vascular cylinder and pericycle. The cortex of roots generally consists of parenchyma cells with large intercellular air spaces. The endodermis (the innermost layer of the cortex adjacent to the pericycle) is composed of closely packed cells that have within their walls Casparian strips, water-impermeable deposits of suberin that regulate water and mineral uptake by the roots. The cortex is surrounded by the dermal system consisting of a single layer of epidermal cells.

The few variations that occur in root anatomy are mainly found among the monocotyledons. The roots of monocotyledons lack secondary growth. Monocotyledons also generally have a parenchymatous pith in the centre of the vascular cylinder and fibres or sclereids, or both, in the cortex; and extensive well-developed pericyclic fibres. Orchids have a multiple-layered epidermis called a velamen, which consists of nonliving compact cells with lignified strips of secondary walls. These cells provide support, prevent water loss, and assist the plant in absorbing water. When dry the orchid root appears white, and when wet the root appears green because the cells of the velamen absorb water, become translucent, and reveal the green cortical cells.

Stems

The shoot apical meristem and the primary meristems lie at the apex of the shoot and give rise to the primary tissues of the stem. The shoot apical meristem produces leaves and axillary buds exogenously; as a result, the epidermis of stems and leaves is continuous. (In contrast, the lateral roots are produced endogenously, and the dermal system of the lateral roots is discontinuous with that of the parent root.)

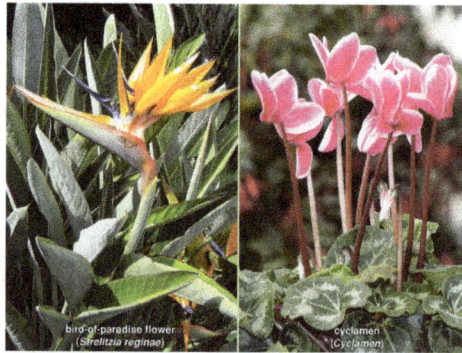

Pedicel; bird-of-paradise flower and Cyclamen.

(Left) The bird-of-paradise flower (Strelitzia reginae) has a long stalk, or pedicel, with large leaves that have prominent midribs (typically red in colour) and long petioles; the leaves, along with the orange sepals and blue petals of the flower, give the plant its bird-like appearance. (Right) Cyclamen grows from tubers, and each flower sits atop a pedicel.

The stem has growth periods similar to those of the root, but longitudinal regions are not as obvious as in the root until the nodes become differentiated and internode lengths increase. Elongation of internodes involves many cell divisions and is followed by cell elongation. At this point, growth in thickness involves some radial cell division and cell enlargement.

The primary tissue systems appear after internode elongation. The procambium differentiates as a basically continuous hollow cylinder or discrete procambial strands, which differentiate into primary xylem and phloem. The ground tissue that lies outside the procambial cylinder is the cortex, and that within is the pith. Ground tissue called the interfascicular parenchyma lies between the procambial strands and remains continuous with the cortex and pith. As the vascular tissue grows, xylem and phloem develop, the vascular bundles mature, the single-layered epidermis differentiates as epidermal cells, trichomes, and a few stomata, and the parenchymatous pith may develop as collenchyma or contain sclereids or fibres or both; unequal pith proliferation and expansion produces the flattened stems (pads) of prickly-pear cacti (Opuntia; Cactaceae). The parenchymatous cortex also may develop some collenchyma, sclereids, or fibres; unequal growth and expansion of the cortex produces the cladodes of epiphytic cacti (e.g., night-blooming cereus, Selenicereus; Cactaceae). In most aquatic angiosperms, the parenchymatous cortex contains large intercellular spaces. As a rule, angiosperm stems have no endodermis or definable pericycle.

The most common arrangement of the primary xylem and phloem is called a collateral bundle; the outer portion of the procambium (adjacent to the cortex) becomes phloem, and the inner portion (adjacent to the pith) becomes xylem. In a bicollateral bundle, the phloem is both outside and inside the xylem, as in Solanaceae (the potato family) and Cucurbitaceae (the cucumber family). In the monocots, the phloem may surround the xylem, or the xylem may surround the phloem.

The vascular bundles of the stem are continuous not only with the primary vascular system of the root but also with the vascular bundles of the leaves. At each node, one or more longitudinal stem bundles enter the base of the leaf as leaf traces, connecting the vascular system of the stem with that of the leaf. The point at which the stem bundle diverges from the vascular cylinder toward the leaf is a leaf gap, called a lacuna. The number of lacunae varies among angiosperm groups and remains a characteristic for classifying the various species.

Several leaves in a line along the stem have common stem bundles. In some species all stem bundles and their associated leaf traces are interconnected, while in others each stem bundle and the associated leaf trace remains laterally independent of the others. An arrangement of two trace leaves and a single lacuna is found among several primitive angiosperm families and throughout the gymnosperms and is the organization from which other nodal patterns are derived.

In woody dicots, the vascular cambium is formed in parts that grow toward each other and unite. Each vascular bundle develops a meristematic area of growth from an undifferentiated (parenchymatous) layer of cells between the primary xylem and primary phloem, called a fascicular cambium. This meristematic area spreads laterally from each bundle and eventually becomes continuous, forming a complete vascular cambium.

The arborescent (treelike) stems of monocotyledons have a different growth pattern and anatomy from dicotyledons. Scattered throughout the ground tissue are vascular bundles with no fascicular cambia and no definable pith or cortex. Secondary growth, when it occurs, is different because a secondary thickening meristem forms under the epidermis. This secondary thickening meristem produces secondary parenchyma (conjunctive tissue) to the inside, and then secondary vascular bundles develop within this conjunctive tissue. Thus, there are no rings of secondary xylem or secondary phloem as in woody dicotyledons.

Many arborescent monocots have only massive primary growth without secondary growth. This primary growth is derived from a primary-thickening meristem under the leaf bases that is a lateral continuation of the apical meristem. This primary-thickening meristem produces vast amounts of parenchyma to the inside, through which the leaf traces differentiate.

Leaves

Leaves initially arise from cell divisions in the shoot apical meristem. A slight bulge (a leaf buttress) is produced, which in dicots continues to grow and elongate to form a leaf primordium. (Stipules, if present, appear as two small protuberances.) Marginal and submarginal meristems on opposite flanks of the primordium initiate leaf-blade formation. Differences in the local activity of marginal meristems cause the lobed shapes of simple leaves and the leaflets in compound leaves. An increase in width and in the number of cell layers is brought about by marginal meristems. Subsequent expansion and increase in length is achieved by cell division and the general enlargement of cells throughout the blade.

Leaf growth is determinate; the tip matures first, and maturation then progresses toward the base, after which the leaf cells cease to grow and divide. (Stem growth is generally indeterminate since the meristems are active indefinitely.) The petiole, when present, and the leaf base become thickened, and often the latter expands laterally and fully or partially encloses the stem. Soon after the cells of the marginal meristems begin to divide, procambial strands differentiate into the leaf from the stem bundles to form the midvein, or midrib. The smaller lateral veins of the leaf are initiated near the leaf tip; subsequent major lateral veins are initiated sequentially toward the base, following the overall pattern of leaf development. A major lateral vein may have one or more orders of smaller veins, which also are initiated in size from larger to smaller. This results in the netlike venation patterns characteristic of dicotyledonous leaves.

The anatomy of a mature dicot leaf generally reflects the habitat, especially the availability of water. Mesomorphic leaves are adapted to conditions of abundant water and relatively humid conditions; xeromorphic leaves are adapted to dry conditions with relatively low humidity; and hydromorphic leaves are adapted to aquatic situations, either submerged or in standing water. Mesomorphic leaves (the most common type) are characteristic of crop plants, such as tomatoes and soybeans. Their veins (vascular bundles) permeate the ground tissue of the dermal system—a single layer of epidermal cells with interspersed guard cells. The ground tissue system, the mesophyll, is divided into two regions: the palisade parenchyma, located beneath the upper epidermis and composed of columnar cells oriented perpendicular to the leaf surface, and spongy parenchyma, located in the lower part of the leaf and composed of irregularly shaped cells. The veins contain primary xylem and phloem and are enclosed by a layer of parenchyma called the bundle sheath. Only the midvein and some large lateral veins have any secondary growth.

The anatomy of mesomorphic leaves is designed to function optimally for water uptake and gas exchange in photosynthesis under mesic (moist) conditions. The spongy mesophyll with irregularly shaped cells provides increased surface area internally, while the elongate palisade cells provide optimal exposure of chloroplasts to light.

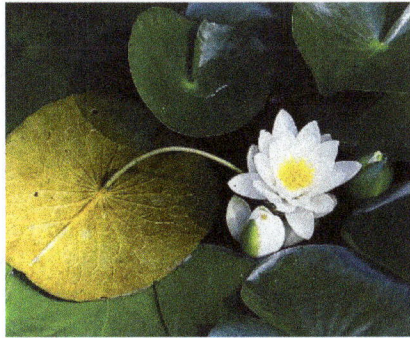

The floating leaf of a water lily facing downward to show the attachment of the leaf stalk near the centre of the leaf. Hydromorphic leaves are thin, and the vascular tissues are scant, because the surrounding water provides mechanical support for the plant.

The anatomy of hydromorphic leaves is simplified: the cuticle is thin or lost; the guard cells are raised and are found only on the upper surface in floating leaves (they are lost in most submerged leaves); the mesophyll contains aerenchyma (an adaptation to promote water loss) and little or no collenchyma or sclerenchyma; and the vascular system (particularly the water-conducting element of the vascular system, the xylem) is only weakly developed since the water provides much of the mechanical support to the plant normally provided by the xylem. The abundance of water means that there is no need for mechanisms that prevent water loss and little need for additional supports. The leaves generally become large and thin, and the reduction or loss of cuticle, vascular tissue, and ground tissue (mesophyll) permits the rapid loss of water vapour (transpiration). The guard cells on the upper surface of floating leaves also monitor the rate of water loss through the central stomata. Such plants may wilt if the turgor (water) pressure is reduced. Water lilies and rice crops contain hydromorphic leaves.

Xeromorphic adaptations to arid conditions are quite varied and tend to prevent water loss during periods when water is limited and must be conserved by the plant. There are many modifications limiting transpiration: two examples are a multilayered epidermis covered by thick layers of epicuticular wax or mucilages secreted into stomates; another is dense mats of trichomes on both surfaces of the leaf and guard cells and stomata sunken into the lower surface and often lined with numerous trichomes, which trap moisture, thereby inhibiting total water loss. Mesophyll modifications provide a means of storing water. Most xeromorphic leaves have a high volume-to-surface ratio—i.e., they are small and compact. In addition, many are fleshy and often oval to round in shape.

The development of monocotyledonous leaves after initiation of cell division on the shoot apical meristem is different from that of dicotyledons and results in leaves with different morphologies from those of dicotyledons. Leaves in monocotyledons have either a radial leaf tip or are expanded in the same plane as the stems instead of at a right angle to the stem, as in dicotyledons. The leaf buttress begins as a ring that encloses the stem. The upper portion of the buttress develops a meristem on the side facing the stem (adaxial meristem). Growth at this adaxial meristem forms the flattened leaf with the radial (cylindrical) leaf tip typical of the monocotyledons. If the adaxial meristem is long-lived, long flat leaves in the same plane as the stem are formed (Iris; Iridaceae); if short-lived , flat leaves with short cylindrical tips develop (snake plant, Sansevieria trifasciata; Agavaceae). When the radial (topmost aspect of the leaf) is short, the base becomes flattened because the marginal meristems (those on either side of the midvein) continue to expand outward. A monocot leaf grows either radially or along the margins, but not both in the same region. The monocot leaf grows in length from a meristem at its base, which is why it is possible to mow grass and have the leaf blades continue to grow.

The developmental pattern from a basal intercalary meristem has placed constraints on the anatomy of monocot leaves, particularly with respect to venation and the position of stomates. This has produced a leaf anatomy characteristic of the monocots. There is no midvein, and veins are longitudinally parallel. The stomates are in rows between the veins, and the mesophyll is often poorly developed and mostly parenchymatous with scattered bundles of fibres. Thus, most monocot leaves are uniform in appearance and texture. Most of the hydromorphic and xeromorphic modifications found among dicot leaves, however, also occur in monocot leaves in similar environments.

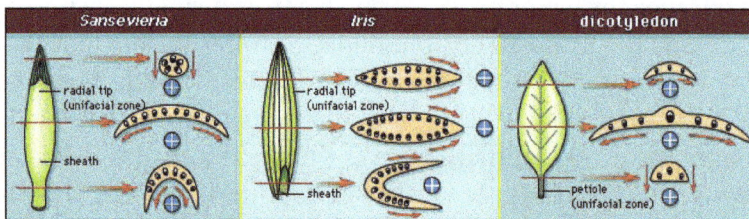

Transections of various leaf types showing principal direction of development.

Reproductive Structures

Features

The broad range of variation in the morphology and structure of nonreproductive (vegetative) organs within the angiosperms has been outlined above. There is a similarly broad range in the morphology and structure of the reproductive organs of the plant.

Many vegetative buds sooner or later become flower buds. Flower buds are modified leaves borne on a short axis with very short internodes and no axillary buds. The floral axis has determinate growth, in that at some point it ceases to grow.

Flowers, the reproductive tissues of the plant, contain the male and female organs. They may terminate short lateral branches or the main axis or both. Flowers may be borne singly (as in the daffodil and Magnolia) or in clusters called inflorescences (e.g., bromeliads, snapdragons, and sunflowers). Fruits are derived from the floral parts of the angiospermous plant.

Rhododendrons in bloom along a trail.

Arum

Titan arum, or corpse flower (Amorphophallus titanum), featuring the largest un-branched inflorescence in the world.

A complete flower is composed of four organs attached to the floral stalk by a receptacle. From the base of the receptacle upward these four organs are the sepals, petals, stamens, and carpels. In dicots the organs are generally grouped in multiples of four or five (rarely in threes), and in monocots they are grouped in multiples of three.

Floral structures characteristic of angiosperms.

The sepals, the outermost layer, are usually green, enclose the flower bud, and collectively are called the calyx. Petals are the next layer of floral appendages internal to the calyx; they are generally brightly coloured and collectively are called the corolla. The calyx and corolla together compose the perianth. The sepals and petals are accessory parts or sterile appendages; though they protect the flower buds and attract pollinators, they are not directly involved with sexual reproduction. When the colour and appearance of sepals and petals are similar, as in the tulip tree (Liriodendron tulipifera) and Easter lily (Lilium longiflorum), the perianth is said to be composed of tepals.

Internal to the corolla are the stamens, spore-producing structures (microsporophylls) that are collectively called the androecium. In most angiosperms, the stamens consist of a slender stalk (the filament) that bears the anther (and pollen sacs), within which the pollen is formed. Small secretory structures called nectaries are often found at the base of the stamens and provide food rewards for pollinators. In some cases the nectaries coalesce into a nectary or staminal disc. In many cases the staminal disc forms when a whorl of stamens is reduced into a nectiferous disc, and in other cases the staminal disc is actually derived from nectary-producing tissue of the receptacle.

At the centre of the flower are the carpels, collectively called the gynoecium. Carpels are megasporophylls that enclose one or more ovules, each with an egg. After fertilization, the ovule matures into a seed, and the carpel matures into a fruit. Carpels, and thus fruit, are unique to angiosperms.

A complete flower contains all four organs, while an incomplete flower is missing at least one. A bisexual (or "perfect") flower has both stamens and carpels, and a unisexual (or "imperfect") flower either lacks stamens (and is called carpellate) or lacks carpels (and is called staminate). Species with both staminate flowers and carpellate flowers on the same plant (e.g., corn) are monoecious, from the Greek for "one house." Species in which the staminate flowers are on one plant and the carpellate flowers are on another are dioecious, from the Greek for "two houses."

Floral organs are often united or fused: connation is the fusion of similar organs—e.g., the fused petals in the morning glory; adnation is the fusion of different organs—for example, the stamens fused to petals in the mint family (Lamiaceae). The basic floral pattern consists of alternating whorls of organs positioned concentrically: from outside inward, sepals, petals, stamens, and carpels. It is possible in most cases to interpret the flower with respect to missing parts and/or modification of parts to function as missing parts simply by positional relationships. In a complete five-merous flower (starting from the outside) there would be a whorl of five sepals, followed by an alternating whorl of five petals, followed by an alternating set of five stamens. In the floral diagram, the midline of each petal is midway between the midlines of two adjacent sepals. Because the whorls alternate, the midline of each stamen of the stamen whorl is between the midlines of two adjacent petals and on the midline of each sepal. When the petals are missing and bracts appear coloured and petaloid as in the Bougainvillea, one of the three whorls is missing: there are only two whorls of five organs instead of the three whorls of five organs described above. Because one whorl of the flower is obviously composed of stamens that bear functional pollen and the other whorl is composed of a brightly coloured set of organs that resemble petals one might conclude that the sepals are missing. But examination of positional relationships between the whorls reveals that the midline of each stamen is on the same line as the midline of the organs of the brightly coloured set. Thus, position tells us that the brightly coloured whorl represents a sepal whorl and that the sepals have assumed the function of the missing petals.

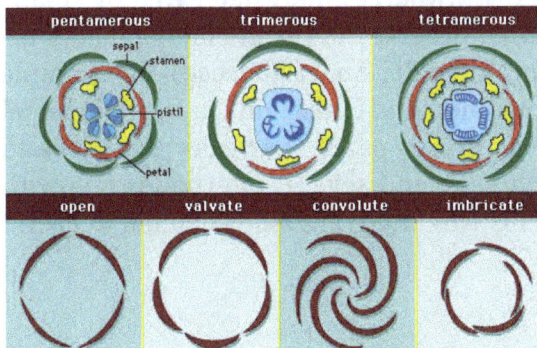

Arrangement of floral parts. (Top) Floral diagrams showing different arrangements of flower structures. (Bottom) Types of arrangement (aestivation) of sepals and petals in flower buds.

The Receptacle

The receptacle is the axis (stem) to which the floral organs are attached. Floral organs are attached either in a low continuous spiral, as is common among primitive angiosperms, or in alternating successive whorls, as is found among most angiosperms.

The peduncle is the stalk of a flower or an inflorescence. When a flower is borne singly, the internode between the receptacle and the bract (the last leaf, often modified and usually smaller than the other leaves) is the peduncle. When the flowers are borne in an inflorescence, the peduncle is the internode between the bract and the inflorescence; the internode between the receptacle of each flower and its underlying bracteole is called a pedicel. Thus, in inflorescences, bracteole is the equivalent of bract, and pedicel is the equivalent of peduncle.

Bracts of the bougainvillea (Bougainvillea). Each cluster of three small tubular flowers is surrounded by colourful petallike bracts.

Often the bract subtending an inflorescence is brightly coloured, as in the poinsettia (Euphorbia pulcherrima; Euphorbiaceae), or provides protection, as in the woody, boat-shaped bracts in many palms. Bracteoles in the inflorescence of Bougainvillea also are brightly coloured to attract pollinators. In some angiosperms, the receptacle becomes fleshy; in the strawberry, for example, the receptacle is the fleshy edible part of the strawberry and, when eaten by small mammals and birds, aids in seed dispersal. In others, the peduncle or pedicel becomes fleshy; in the cashew (Anacardium occidentale; Anacardiaceae), for example, the pedicel is made into a drink in the Neotropics, and it also aids in fruit dispersal of the much smaller cashew nut. In cacti (e.g., prickly pear), the fleshy part of the edible fruit forms from the receptacle and peduncle, and several internodes below that grow up and surround the carpels; this is why there are axillary buds in cacti (areoles) with spines on the fruit surface.

The Calyx

The sepals (collectively called the calyx) most resemble leaves because of their generally green colour. From their base and along most of their length, sepals remain either separate (aposepalous, or polysepalous) or marginally fused (synsepalous), forming a tube with terminal lobes or teeth. The number of calyx lobes equals the number of fused (connate) sepals.

The sepals enclose and protect the unopened flower bud. The calyx is commonly persistent and evident when the fruit matures (e.g., persimmon, Diospyros virginiana; Ebenaceae), in contrast to the more short-lived petals and stamens. Sepals may be brightly coloured and function as petals when true petals are missing—for example, the virgin's bower (Clematis; Ranunculaceae) and the Bougainvillea. Petaloid sepals in this case differ from tepals because the first group of stamens are on the same radii as the sepals, indicating the absence of the petals, which would normally be positioned on alternating radii in the next floral whorl.

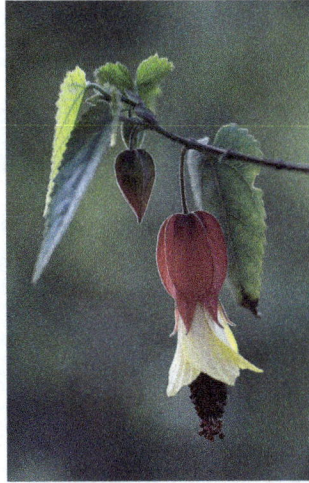

A conspicuous red calyx tube envelops the closed yellow petals of the bell-shaped Abutilon megapotamicum. The anthers are attached to the apex of the exserted staminal column.

The Corolla

The petals composing the corolla are typically brightly coloured or white and attract insects and birds for pollination . The number of petals is usually the same as the number of sepals. Floral symmetry is defined by the petals. When the petals of the corolla are of the same size and shape and when they are equidistant from each other, the flower has radial symmetry, and the flower is called regular or actinomorphic (e.g., buttercup, Ranunculus; Ranunculaceae). In regular flowers, any line drawn through the centre will divide the flower into two identical halves. When at least one petal of the corolla is different, the flower has bilateral symmetry and is called irregular or zygomorphic (e.g., violets, Viola; Violaceae).

Representative forms of the corolla.

Zygomorphy, or bilateral symmetry, of the viola (Viola), which produces a delicate five-petaled flower with two dissimilar pairs. Nectar guides are prominent on the lower spurred petal.

The petals of the corolla may be separate, or apopetalous, or marginally fused (fusion of like floral parts is called connation), or sympetalous, for all or part of their length. When joined, they form a tubular corolla with terminal lobes. A tubular corolla may be present in regular flowers (e.g., blueberries, Vaccinium; Ericaceae) or irregular flowers (e.g., sage, Salvia officinalis; Lamiaceae). Stamens are commonly united to a tubular corolla (fusion of two unlike floral parts is called adnation). A marginally fused (synsepalous) calyx, a marginally fused (sympetalous) corolla, and stamens may fuse to form a cuplike floral tube called a hypanthium that surrounds the carpels, as in cherries (Prunus; Rosaceae), for example. Fusion and reduction of flower parts are common and have occurred in many unrelated lineages. Many wind-pollinated angiosperms do not have petals, nor do they have floral parts modified as petals; examples of wind-pollinated species include the amaranth family (Amaranthaceae) and the birch family (Betulaceae).

The petals of the delicate flower of Campsis radicans (trumpet creeper, or trumpet vine) form a corolla tube with five spreading lobes. A shortened calyx tube covers the base of the flower.

Petals often bear nectaries that secrete sugar-containing compounds, and petals also produce fragrances to attract pollinators; the fragrance of a rose (Rosa; Rosaceae) is derived from the petals. Petals often develop a nectar-containing extension of the tubular corolla, called a spur. This may involve one petal, as in the larkspur (Delphinium), or all the petals, as in columbine (Aquilegia), both members of the family Ranunculaceae.

The Androecium

Stamens (microsporophylls) are structures that produce pollen in terminal saclike structures (microsporangia) called anthers. The number of stamens comprised by the androecium is sometimes the same as the number of petals, but often the stamens are more numerous or fewer in number than the petals. There are generally two pairs of spore-containing sacs (microsporangia) in a young stamen; during maturation the partition between the adjacent microsporangia of a pair breaks down so that there are only two pollen-containing sacs (one in each lobe of the anther) at the time the stamen releases the pollen.

The least-modified stamens are similar to leaves, with the paired microsporangia located at or near the margins; an example is found in the magnolia family (Magnoliaceae). In more derived stamens, the blade has become modified into a slender stalk, the filament, with the microsporangia at or near the filament apex (the anther). The filaments are very often united with the corolla, but with the anthers either separate, as in primroses (Primula; Primulaceae), or united with each other to form a staminal tube that encloses the gynoecium, as in the mallow family (Malvaceae). In thistle (Cirsium; Asteraceae) and in other members of the sunflower family, the staminal tube is fused to the lower half of the corolla tube.

There are several trends in stamen modification. In many angiosperms, one or more of the stamens is modified and lacks functional anthers. In the most common modification, the filament is expanded to form a petallike blade called a staminode (in the same manner that a sepal forms a petallike blade in some flowers without true petals). The apparent petals in some angiosperm families, such as are found in many members of the pink family (Caryophyllaceae), are staminodial in origin. Wild roses have only five petals and many stamens; however, cultivated roses have been selected for the many apparent petals (but actually staminodes) and few functional stamens. In other cases, stamens have been modified into sterile nectaries involved in pollination. If flowers have a large number of stamens, then the stamens often occur in groups or clusters, as in the myrtle family (Myrtaceae).

The brilliant regular flower of Hypericum calycinum (rose of Sharon) develops a superior ovary with five spreading styles at its apex and numerous stamens arranged in five clusters (fascicles) emanating from below the base of the ovary.

The Gynoecium

The gynoecium is composed of carpels. In more basal families (e.g., Magnoliaceae) the carpels are spirally arranged, and in more advanced families they tend to be arranged in a single whorl. Carpel number varies from one (e.g., bean or legume family [Fabaceae]) to many (e.g., buttercups or raspberries [Rubus]).

At the base of a carpel is the ovary, within which develop one or more multicellular structures called ovules that each contain an egg. The upper part of the carpel, the stigma, receives the pollen. A slender stalk called the style often connects the ovary and stigma. The carpels may be separate (apocarpous) or fused together (syncarpous), with the individual carpel walls and cavities (locules) still present. Syncarpy may involve only the ovaries, leaving the styles and stigmas free, as is found in the wood sorrel (Oxalis), or it may involve both the ovaries and styles, leaving only the stigmas free, as in the waterleaf (Hydrophyllum). The number of carpels in a syncarpous (or compound) ovary generally equals the number of locules (in some cases the inner carpel walls break down, leaving a single locule); in an orange or a grapefruit, for example, the juice sacs are actually trichomes that line the inner carpel walls of each cavity.

The position of the gynoecium with respect to the petals, sepals, and stamens on the floral axis also characterizes the flower. In hypogynous flowers, the perianth and stamens are attached to the receptacle below the gynoecium; the ovary is superior to these organs, and the remaining floral organs arise from below the point of origin of the carpel. In perigynous flowers, a hypanthium (a floral tube formed from the fusion of the stamens, petals, and sepals) is attached to the receptacle below the gynoecium and surrounds the ovary; the ovary is superior, and the free parts of the petals, sepals, and stamens are attached to the rim of the hypanthium. In epigynous flowers, the hypanthium is fused to the gynoecium, and the free parts of the sepals, petals, and stamens appear to be attached to the top of the gynoecium, as in the apple (Malus; Rosaceae); the ovary is inferior, and the petals, sepals, and stamens appear to arise from the top of the ovary.

Fruits

Fertilization of an egg within a carpel by a compatible pollen grain results in seed development within the carpel. (Formation of fruit without the fertilization of an egg and subsequent seed development is called parthenocarpy.) A fruit is a ripened ovary (or compound ovary) and any other structure, usually the hypanthium, that ripens and forms a unit with it. This clearly separates a fruit from a vegetable, because a vegetable is derived only from vegetative (nonreproductive) organs. Tomatoes, eggplants, and squashes are fruits, because they are derived from floral parts, whereas carrots, turnips (Brassica rapa), and beets are vegetables, because they are roots modified as storage organs in the same manner that potatoes, ginger (Zingiber officinale), and onions are modified stems.

Simple fruits develop from a single carpel or from a compound ovary. Aggregate fruits consist of several separate carpels of one apocarpous gynoecium (e.g., raspberries where each unit is a single carpel). Multiple fruits consist of the gynoecia of more than one flower and represent a whole inflorescence, such as the fig and pineapple. Accessory fruits incorporate other flower parts in the development of the mature fruit; for example, the hypanthium is used in forming the pear (Pyrus; Rosaceae), and the receptacle becomes part of the prickly pear.

The form, texture, and structure of fruits are varied (notably in simple fruits), but most fall within a few categories. The fruit wall, or pericarp, is divided into three regions: the inner layer, or endocarp; the middle layer, or mesocarp; and the outer layer, or exocarp. These regions may be fleshy or dry (sclerified) or any combination of the two, but they are classified as either one or the other.

The three main types of fleshy fruits are berries, drupes, and pomes. Berries are many-seeded simple fruits composed of one carpel or a syncarpous ovary. They are fleshy throughout, but the exocarp ranges in texture: a soft, thin exocarp, as in tomatoes (a berry); a leathery exocarp, as in oranges (a hesperidium); and a somewhat hard exocarp, as in pumpkins (a pepo). In drupes, or stone fruits, there is usually only one seed per carpel or locule. Drupes are fleshy fruits and consist of an inner stony or woody endocarp, which adheres to the seed (peaches, plums, and cherries). The term druplet is used for each unit of aggregate fruit of this type (e.g., raspberries and blackberries). Pomes are fleshy fruits of the rose family (Rosaceae) in which an adnate hypanthium becomes fleshy (apples and pears).

Simple dry fruits are either dehiscent or indehiscent. They are dehiscent if the pericarp splits open at maturity and releases the seeds, or indehiscent if the pericarp remains intact when the fruit is shed from the plant. The three principal types of dehiscent fruits are follicles, legumes, and capsules. Follicles and legumes are each derived from an ovary with a single carpel, and a capsule is derived from several united carpels. As the fruit matures, the pericarp dries and the fruit splits. Whereas follicles split along a single side of the fruit, such as in the milkweeds, columbines, and magnolias, legumes split along both sides, as in the bean family. Capsules have two or more carpels and split open to release their seeds in various ways. They may open longitudinally to expose the seeds within each locule (cavity) or longitudinally along each septum between the locules, as in the agave (Agave; Agavaceae). Still others form an operculum (a lid) at the top of the ovary, as in the Brazil nut family (Lecythidaceae).

Indehiscent fruits are derived from either single carpels or compound ovaries. Single carpel forms include the achene, the samara, and the caryopsis. Forms derived from a compound ovary include nuts and schizocarps. An achene is a fruit in which the single seed lies free in the cavity, attached only by a single point. The strawberry, for example, is really an aggregate fruit, and each "seed" is an achene. The samara is a winged achene and is found in the tree of heaven (Ailanthus altissima; Simaroubaceae) and ash (Fraxinus; Oleaceae). In the caryopsis, or grain, the seed adheres to the fruit wall (pericarp). The caryopsis is found among the cereal grasses, such as corn. Nuts have a stony pericarp, and usually only a single seed in each carpel matures, as in acorns of oaks (Quercus; Fagaceae) and hazelnuts (Corylus avellana; Betulaceae). Schizocarps are fruits in which each carpel of a compound ovary splits apart to form two or more parts, each with a single seed. Schizocarps are found in the carrot family (Apiaceae). Winged schizocarps are found in maples.

Seeds

Seeds are the mature ovules. They contain the developing embryo and the nutritive tissue for the seedling. Seeds are surrounded by one or two integuments, which develop into a seed coat that is usually hard. They are enclosed in the ovary of a carpel and thus are protected from the elements and predators.

The ovule is attached to the ovary wall until maturity by a short stalk called the funiculus. The area of attachment to the ovary wall is referred to as the placenta. The arrangement of placentae (placentation) in the compound ovary of angiosperms is characterized by the presence or absence of a central column in the ovary and by the site of attachment. In axile placentation the placentae are located on a central column; partitions from the central column to the ovary wall create chambers (locules) that separate the placentae and attached ovaries from each other. Free-central placentation resembles axile placentation; however, the column is not connected by partitions to the ovary wall, and thus no locules are formed. In basal placentation ovules are attached to the base of the ovary, and in parietal placentation the placentae are located directly on the ovary wall or on its extensions.

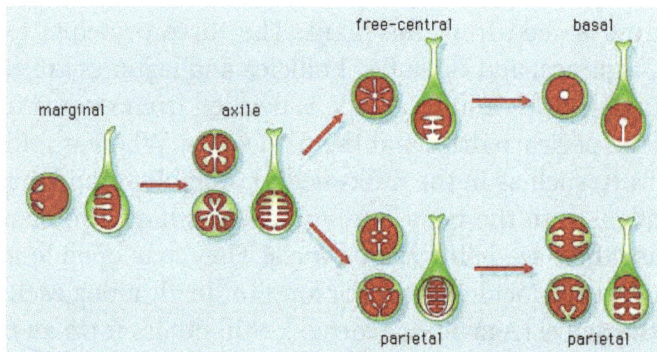

Evolutionary relationships among some types of placentation.

Mature seeds are enclosed in integuments that may become hard and stony or that may have an outer fleshy, usually brightly coloured sarcotesta with an inner stony sclerotesta. Seed coats also may be winged or variously ornamented with prickles or sclerified hairs. In some seeds, there may be an extra covering, the aril, which is an outgrowth of the funiculus (e.g., the spice mace is derived from the red aril of Myristica fragrans; Myristicaceae). The aril of tomato seeds makes them slippery.

Mechanisms of Dispersal

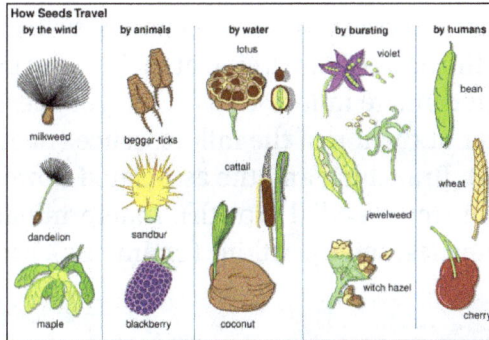

Seed Dispersal Seeds and their dispersal mechanisms.

Fruits and seeds are the primary means by which angiosperms are dispersed. The chief agents of dispersal are wind, water, and animals. Some fruits and seeds have modifications that aid in wind dispersal. Fruit modifications include samaras, samaroid schizocarps, and the feathery calyx lobes (e.g., dandelion). Seeds may be modified in various ways to promote dispersal: they may be extremely small and light (e.g., orchids, Orchidaceae), winged (e.g., common catalpa, Catalpa bignonioides; Bignoniaceae), plumed (e.g., milkweed), covered with woolly hairs (e.g., willows), or surrounded by explosive capsules that forcefully eject them into the air, as, for example, the touch-me-not (Impatiens; Balsaminaceae) and the witch hazel (Hamamelis; Hamamelidaceae). The fruits or seeds of many aquatic and shore plants are adapted to float on water as a means of dispersal; for this reason, coconuts (Cocos nucifera; Arecaceae) are readily transported across oceans to neighbouring islands. Adaptations for water dispersal include aerenchyma in fruits or seeds and light weight (e.g., water chestnut, Trapa natans; Lythraceae).

Animals disperse fruits and seeds either by ingesting and subsequently excreting them or by passively transporting them once they have adhered to an external part of the body, such as the fur or a claw. The evolution of fleshy fruits and seeds exemplifies the coevolution of plants and their animal agents of dispersal. An animal diet often consists solely of fruits and seeds that are designed to be eaten and dispersed, and in many cases these seeds require full or partial digestion to stimulate germination. Most fruits with a fleshy pericarp are eaten whole by vertebrates, including the stony endocarp or the stony seed coat. The seeds then either are regurgitated by the animal or pass through the alimentary canal and are excreted, often some distance from the original site. Seeds with an aril that encloses a stony seed coat or seeds with a sclerotesta and a fleshy, coloured sarcotesta are found in dehiscent fruits. They are eaten by animals after the fruit has ripened and split open. Often these seeds dangle from the fruit by long stalks (e.g., the follicles of Magnolia). The fleshy portion, whether originally a fruit or seed, is brightly coloured and sweet so as to attract vertebrates, particularly birds and mammals. Many fruits and seeds in the Amazon, however, are actually eaten and dispersed by fish during times of high water.

Inflorescences

Inflorescences are clusters of flowers on a branch or a system of branches. They are categorized generally on the basis of the timing of their flowering and by their arrangement on an axis. In indeterminate inflorescences, the youngest flowers, and therefore the last to open, are either at the top of the inflorescence (in elongated axes) or in the centre (in truncated axes). Branching and the associated flowers develop at some distance from the main stem (monopodial growth). Indeterminate inflorescences are of varied types: racemes, panicles, spikes, catkins (or aments), corymbs, and heads.

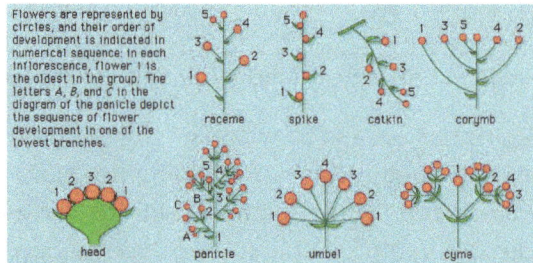

Angiosperm inflorescences: Common types of inflorescences among the angiosperms.

A raceme is an inflorescence in which a flower develops at the axil of each leaf along an elongated, unbranched axis . Each flower terminates a short stalk called a pedicel. The main axis has indeterminate growth; therefore, its growth does not cease at the onset of flowering. A spike is a raceme except that the flowers are attached directly to the axis at the axil of each leaf rather than to a pedicel. An example of a spike is the cattail (Typha; Typhaceae). The fleshy spike characteristic of the Araceae is called a spadix, and the underlying bract is known as a spathe. A catkin (or ament) is a spike in which all the flowers are of only one sex, either staminate or carpellate. The catkin is usually pendulous, and the petals and sepals are reduced to aid in wind pollination when the inflorescence as a whole is shed. An example of a catkin is found in oaks. A corymb is a raceme in which the pedicels of the lower flowers are longer than those of the upper ones so that the appearance of the inflorescence overall is that of a flat flower. The lower flowers open first, and the axis of a corymb continues to produce flowers (indeterminate growth). Corymbs are found in the hawthorn (Crataegus; Rosaceae).

A raceme of lily of the valley (Convallaria majalis).

Spikes of false dragonhead, or obedience plant (Physostegia angustifolia).

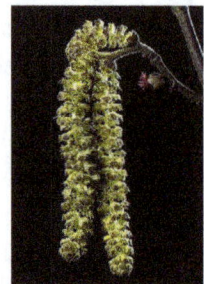

A drooping male catkin (left) and the small red female inflorescence (right) of hazel (Corylus avellana).

Corymbs of yarrow (Achillea taygetea).

If the axis is short or stunted, the flowers arise from a common point and appear to be at approximately the same level. This pattern, called an umbel, is actually a flattened raceme because the internodes of the axis, or peduncle (the point of origin of the leaves and flower axes), are shortened so that the pedicels are of the same length (e.g., the carrot family). A head is a raceme in which the peduncle is flattened and the flowers are attached directly to it (e.g., aster family, Asteraceae). This results in a grouping of small flowers in such a way as to appear as a single flower. In many members of the Asteraceae (e.g., sunflowers, Helianthus annuus), for instance, the outer (or ray) flowers have a well-developed zygomorphic corolla, and the inner (disk) flowers have a small actinomorphic corolla. The inner disk flowers generally are complete flowers, and the ray flowers generally are sterile.

Simple umbels of the Texas, or white, milkweed (Asclepias texana).

In the compound indeterminate inflorescences, the main axis is branched so that the many inflorescences form off the main axis. A panicle is a branched raceme in which the branches are themselves racemes (e.g., yuccas, Yucca). In a compound umbel, all the umbel inflorescences arise from a common point and appear to be at about the same level (e.g., wild carrot). This organization is the same for compound spikes, catkins, corymbs, and heads. The change from elongated axes (racemes and panicles) to flattened axes (corymbs and umbels) results in inflorescences in which the flowers are arranged close together. This close association encourages efficient pollination, and the extreme condensation of the inflorescences, as in the head, gives rise to an inflorescence that appears to be a single flower (e.g., sunflowers).

Panicles of astilbe (Astilbe).

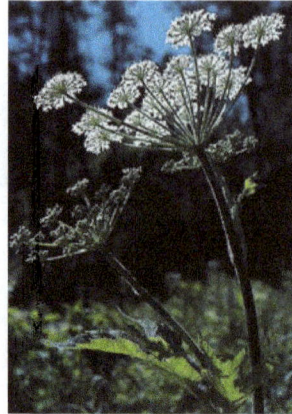
Common hogweed Compound umbels of common hogweed (Heracleum sphondylium).

In the determinate (cymose) inflorescences, the youngest flowers (those that are the last to open) are at the bottom of an elongated axis or on the outside of a truncated axis (e.g., in the cymose umbel of onions, Allium; Alliaceae). These inflorescences are determinate because, at the time of flowering, the whole apical meristem produces a flower; thus, the entire axis ceases to grow. Each unit of a cyme consists of a dichasium, which has a central flower and two lateral flowers. The branching is primarily sympodial, and the inflorescence may be compound (e.g., catchfly, or campion, Silene; Caryophyllaceae). Many monocotyledons have a one-sided cyme called a helicoid cyme A cymose inflorescence arranged in pairs at the nodes, in the manner of a false whorl, is called a verticillaster. Finally, there are mixed inflorescences, as, for instance, the cymose clusters arranged in a racemose manner (e.g., lilac, Syringa vulgaris; Oleaceae) or other types of combinations.

A dichasium (the basic unit of a cyme) of the wood stichwort (Stellaria nemorum).

A helicoid cyme of the fiddle-neck (Amsinckia intermedia), a type of determinate inflorescence in which the flowers develop on one side of the axis, causing the inflorescence to curve.

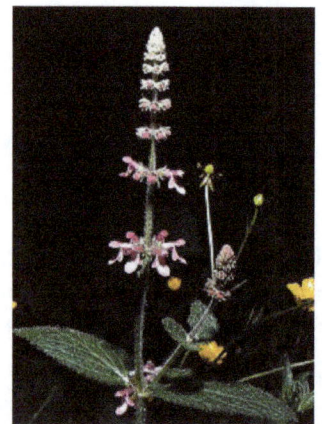
Verticillasters (a false whorl of inflorescences found at the nodes) of the hedge nettle (Stachys bullata).

Reproduction

Features

The vast array of angiosperm floral structures is for sexual reproduction. The angiosperm life cycle consists of a sporophyte phase and a gametophyte phase. The cells of a sporophyte body have a full complement of chromosomes (i.e., the cells are diploid, or 2n); the sporophyte is the typical plant body that we see when we look at an angiosperm. The gametophyte arises when cells of the sporophyte, in preparation for reproduction, undergo meiotic division and produce reproductive cells that have only half the number of chromosomes (i.e., haploid, or n). A two-celled microgametophyte called a pollen grain germinates into a pollen tube and through division produces the haploid sperm. (The prefix micro- denotes gametophytes emanating from a male reproductive organ.) An eight-celled megagametophyte called the embryo sac produces the egg. (The prefix mega- denotes gametophytes emanating from female reproductive organs).

Angiosperms are vascular plants, and all vascular plants have a life cycle in which the sporophyte phase (vegetative body) is the dominant phase and the gametophyte phase remains diminutive. In the nonvascular plants, such as the bryophytes, the gametophyte phase is dominant over the sporophyte phase. In bryophytes, the gametophyte produces its food by photosynthesis (is autotrophic) while the nongreen sporophyte is dependent on the food produced by the gametophyte. In nonseed vascular plants, such as ferns and horsetails, both the gametophyte and sporophyte are green and photosynthetic, and the gametophyte is small and without vascular tissue. In the seed plants (gymnosperms and angiosperms), the sporophyte is green and photosynthetic and the gametophyte depends on the sporophyte for nourishment. Within the seed plants, the gametophyte has become further reduced, with fewer cells comprising the gametophyte. The microgametophyte (pollen grain), therefore, is reduced from between 4 and 8 cells in the gymnosperms to a 3-celled microgametophyte in the angiosperms. A parallel reduction in the number of cells comprising a megagametophyte (ovule) has also taken place: from between 256 and several thousand cells in the gymnosperms to an 8-celled megagametophyte in most of the angiosperms. The significance of the reduction in megagametophyte cells appears to be related to pollination and fertilization. In many gymnosperms, pollination leads to the formation of a large gametophyte with copious amounts of stored starch for the nourishment of the potential embryo regardless of whether fertilization of the ovule can actually take place (i.e., whether the pollen is from the same species as the ovule). If the pollen is from a different species, fertilization or embryo development fails, so that the stored food is wasted. In angiosperms, however, the megagametophyte and egg are mature before the food is stored, and this is not ever accomplished until after the egg has been adequately fertilized and an embryo is present. This reduces the chances that the stored food will be wasted.

The process of sexual reproduction depends on pollination to bring these gametophytes in close association so that fertilization can take place. Pollination is the process by which pollen that has been produced in the anthers is received by the stigma of the ovary. Fertilization occurs with the fusion of a sperm with an egg to produce a zygote, which eventually develops into an embryo. After fertilization, the ovule develops into a seed, and the ovary develops into a fruit.

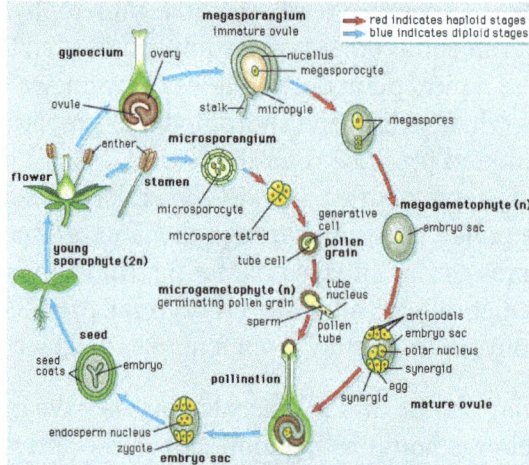

Typical angiosperm life cycle

Anthers

A transverse section of the anther reveals four areas of tissue capable of producing spores. These tissues are composed of microsporocytes, which are diploid cells capable of undergoing meiosis to form a tetrad (four joined cells) of haploid microspores. The microspores become pollen grains and may eventually separate.

During pollen development, the layer of cells beneath the dermis of the anther wall (the endothecium) develops thickenings in the cell walls. The cell layer immediately inside the endothecium (the tapetum) develops into a layer of nutritive cells that either secrete their contents into the area around the microsporocytes or lose their inner cell walls, dissociate from each other, and become amoeboid among the microsporocytes. The pollen grains develop a thick wall of at least two layers, the intine and the exine. The intine, or inner layer, consists primarily of cellulose and pectins. The exine, or outer layer, is composed of a highly decay-resistant chemical called sporopollenin. The exine usually has one or more thin areas, or pores, through which the pollen tubes germinate, and the thick area of the exine is usually highly sculptured. The number of pores and pattern of exine sculpturing are characteristic within an angiosperm family, genus, and often within a species.

The terminology to describe the various sculpturing patterns and position and number of pores is highly complex and only a basic description as related to functional aspects of sculpturing is given here. For example, smooth or essentially smooth pollen is loosely correlated with wind pollination, as in oaks (Quercus) and grasses (corn, Zea mays). Many plants pollinated by birds, insects, and small mammals have highly sculptured patterns of spines, hooks, or sticky threadlike projections by which pollen adheres to the body of the foraging pollinator as it travels to other flowers.

Each microspore (pollen grain) divides mitotically to form a two-celled microgametophyte; one cell is a tube cell (the cell that develops into a pollen tube), and the other is a generative cell, which will give rise to two sperm as a result of a further mitotic division. Thus, a mature microgametophyte consists of only three haploid cells—the tube cell and two sperm. Most angiosperms shed pollen at the two-celled stage, but in some advanced cases it is shed at the mature three-celled stage. When the pollen grains are mature, the anther wall either splits open (dehisces) longitudinally or opens by an apical pore.

Because the sporopollenin is resistant to decay, free pollen is well represented in the fossil record. The distinctive patterns of the exine are useful for identifying which species were present as well as suggesting the conditions of early climates. The proteins in the pollen walls are also a major factor in hay fever and other allergic reactions, and the spinose sculpturing patterns may cause physical irritation.

Ovules

An ovule is a saclike structure that produces the megaspores and is enclosed by layers of cells. This megasporangium is called the nucellus in angiosperms. After initiation of the carpel wall, one or two integuments arise near the base of the ovule primordium, grow in a rimlike fashion, and enclose the nucellus, leaving only a small opening called the micropyle at the top. In angiosperms the presence of two integuments is plesiomorphic (unspecialized), and one integument is apomorphic (derived). A single large megasporocyte arises within the nucellus near the micropyle and undergoes meiotic division, resulting in a single linear tetrad of megaspores. Three of the four megaspores degenerate, and the surviving one enlarges. The resulting megagametophyte produces the female gametes (eggs). This development (called megagametogenesis) involves free-nuclear mitotic divisions. The cell wall remains intact while the nucleus divides until the megagametophyte, or embryo sac, is formed. The embryo sac typically has eight nuclei. Free-nuclear mitotic division is also found in gametophyte formation in gymnosperms.

Four nuclei migrate to either end of the embryo sac. One nucleus from each group then migrates to the centre of the embryo; these become the polar nuclei. The two polar nuclei merge to form a fusion nucleus in the centre of the embryo sac. A cell wall develops around the fusion nucleus, leaving a central cell in the sac. Cell walls form around each of the chalazal nuclei to form three antipodal cells. During development, enlargement of the embryo sac leads to the destruction of most of the nucellus. This sequence of megasporogenesis and megagametogenesis, called the Polygonum type, occurs in 70 percent of the angiosperms in which the life cycle has been charted. Variations found in the remaining 30 percent represent derivations from the Polygonum type of seed development.

Pollination

Effective pollination involves the transfer of pollen from the anthers to a stigma of the same species and subsequent germination and growth of the pollen tube to the micropyle of the ovule.

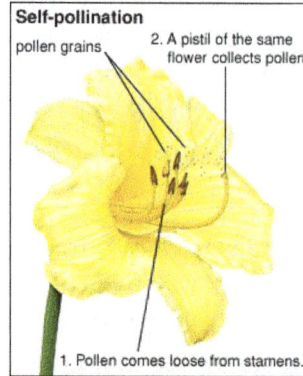

Self-pollination The process of self-pollination in an angiosperm.

Cross-pollination The process of cross-pollination using an animal pollinator.

Pollen transfer is effected by wind, water, and animals, primarily insects and birds. Wind-pollinated flowers usually have an inconspicuous reduced perianth, long slender filaments and styles, covered with sticky trichomes and often branched stigmas, pendulous catkin inflorescences, and small, smooth pollen grains.

Spikes of sedge (Carex pendula) showing reduced floral parts adapted to wind pollination. The pollen bursts forth from the pendulous inflorescences as they sway in the wind.

Wind pollination is derived in angiosperms and has developed independently in several different groups. For example, within the aster family wind pollination accompanied by floral reduction has developed independently in the tribes Heliantheae and Anthemideae. Water pollination occurs in only a few aquatic plants and is highly complicated and derived.

There is a wide range of animal pollinators of angiosperms as well as a wide range of adaptations by the flowers to attract those pollinators. Some of the living unspecialized families of basal angiosperms are pollinated by beetles. The beetles forage and feed on pieces of the perianth and stamens. There are no nectaries but rather food bodies on these organs.

Bees are responsible for the pollination of more flowers than any other animal group. Bees usually feed on nectar and in some cases on pollen. They may be general pollinators by visiting flowers of many species, or they may have adapted (i.e., elongated) their mouthparts to different flower depths and have become specialized to pollinate only a single species. Flowers pollinated by bees commonly have a zygomorphic, or bilaterally symmetrical, corolla with a lower lip providing a landing platform for the bee. Nectar is commonly produced either at the base of the corolla tube or in extensions of the corolla base. The bees partially enter the corolla mouth to feed with their long tongues on the nectar, at which point they deposit pollen picked up from other flowers and collect pollen from the new flower. Flowers pollinated by bees are often blue or yellow or exhibit patterns of both. Particular pattern markings and ultraviolet reflection patterns serve as recognition guides.

An evening primrose (Oenothera biennis) seen (top) in visible light and (bottom) in ultraviolet light; the latter reveals nectar-guide patterns that are discernible to the moth pollinating this flower but not to the human eye.

A high degree of coevolution is common in orchids (e.g., Ophrys speculum), where the flower not only appears to resemble the female wasp of a particular species but also produces the pheromone released by the insect to attract males of the species. The male wasp effects pollination by pseudocopulation with the orchid flower. Other insect pollinators include flies, butterflies, moths, and mosquitoes. Many flowers pollinated by flies are called carrion flowers because they look and smell like rotting meat. The skunk cabbage (Symplocarpus foetidus) and the carrion flowers (Stapelia schinzii) have evolved these characteristics independently.

The labellum of the mirror ophrys (Ophrys speculum). The colouring so closely resembles that of the female wasp Colpa aurea that males of the species are attracted to the flower and pick up pollen during their attempts at copulation.

Orange-tailed butterfly (Eurema proterpia) on an ash-coloured aster (Machaeranthera tephrodes). The upstanding yellow stamens are tipped with pollen, which brushes the body of the butterfly as it approaches the centre of the flat-topped aster to feed on the nectar.

Vertebrate pollinators include birds, bats, small marsupials, and small rodents. Many bird-pollinated flowers are bright red, especially those pollinated by hummingbirds . Hummingbirds rely solely on nectar as their food source. Flowers (e.g., Fuchsia) pollinated by birds produce copious quantities of nectar but little or no odour because birds have a very poor sense of smell. Flowers pollinated by bats produce large quantities of nectar and strong fragrances. They generally open only at night, when bats are the most active, and often hang down on long inflorescence stalks, which provide easy access to the nectaries and pollen. Some eucalypts (Eucalyptus) are pollinated by small marsupials (e.g., honey possums).

Costa's hummingbird (Calypte costae) foraging for nectar in the bright red tubular flowers of ocotillo (Fouquieria splendens). Pollen is displaced onto the beak and head of the bird as it inserts its long tongue into the corolla tube where the nectar is located.

Whatever the agent of dispersal, the first phase of pollination is successful when a pollen grain lands on a receptive stigma. The surface of the stigma can be wet or dry and is often composed of specialized glandular tissue; the style is lined with secretory transmitting tissue. Their secretions provide an environment that nourishes the pollen tube as it elongates and grows down the style. If mitosis in the generative cell has not yet occurred in the pollen grain, it does so at this time.

To prevent self-fertilization, many angiosperms have developed a chemical system of self-incompatibility. The most common type is sporophytic self-incompatibility, in which the secretions of the stigmatic tissue or the transmitting tissue prevent the germination or growth of incompatible pollen. A second type, gametophytic self-incompatibility, involves the inability of the gametes from the same parent plant to fuse and form a zygote or, if the zygote forms, then it fails to develop. These systems force outcrossing and maintain a wide genetic diversity.

The pollen tube ultimately enters an ovule through the micropyle and penetrates one of the sterile cells on either side of the egg (synergids). These synergids begin to degenerate immediately after pollination. Pollen tubes can reach great lengths, as in corn, where the corn silk consists of the styles for the corn ear and each silk thread contains many pollen tubes.

Fertilization and Embryogenesis

After penetrating the degenerated synergid, the pollen tube releases the two sperm into the embryo sac, where one fuses with the egg and forms a zygote and the other fuses with the two polar nuclei of the central cell and forms a triple fusion, or endosperm, nucleus. This is called double fertilization because the true fertilization (fusion of a sperm with an egg) is accompanied by another fusion process (that of a sperm with the polar nuclei) that resembles fertilization. Double fertilization of this type is unique to angiosperms. The zygote now has a full complement of chromosomes (i.e., it is diploid), and the endosperm nucleus has three chromosomes (triploid). The endosperm nucleus divides mitotically to form the endosperm of the seed, which is a food-storage tissue utilized by the developing embryo and the subsequent germinating seed. It has been shown that some of the most basal angiosperms actually form diploid endosperm, although they still experience double fertilization.

The three principal types of endosperm formation found in angiosperms—nuclear, cellular, and helobial—are classified on the basis of when the cell wall forms. In nuclear endosperm formation, repeated free-nuclear divisions take place; if a cell wall is formed, it will form after free-nuclear division. In cellular endosperm formation, cell-wall formation is coincident with nuclear divisions. In helobial endosperm formation, a cell wall is laid down between the first two nuclei, after which one half develops endosperm along the cellular pattern and the other half along the nuclear pattern. Helobial endosperm is most commonly found in the Alismatales (monocotyledons). In many plants, however, the endosperm degenerates, and food is stored by the embryo (e.g., peanut [groundnut], Arachis hypogaea), the remaining nucellus (known as perisperm; e.g., beet), or even the seed coat (mature integuments). Cellular endosperm is the least specialized type of endosperm with nuclear and helobial types derived from it.

The zygote undergoes a series of mitotic divisions to form a multicellular, undifferentiated embryo. At the micropylar end there develops a basal stalk or suspensor, which disappears after a very short time and has no obvious function in angiosperms. At the chalazal end (the region opposite the micropyle) is the embryo proper. Differentiation of the embryo—e.g., the development of cells and organs with specific functions—involves the development of a primary root apical meristem (or radicle) adjacent to the suspensor from which the root will develop and the development of one cotyledon (in monocotyledons) or two cotyledons (in dicotyledons) at the opposite end from the suspensor. A shoot apical meristem then differentiates between the two cotyledons or next to the single cotyledon and is the site of stem differentiation.

The mature embryo is a miniature plant consisting of a short axis with one or two attached cotyledons. An epicotyl, which extends above the cotyledon(s), is composed of the shoot apex and leaf primordia; a hypocotyl, which is the transition zone between the shoot and root; and the radicle. Angiosperm seed development spans three distinct generations, plus a new entity: the parent sporophyte, the gametophyte, the new sporophyte, and the new innovation—namely, the endosperm.

Seedlings

Mature seeds of most angiosperms pass through a dormant period before eventually developing into a plant. The life span of angiosperm seeds varies from just a few days (e.g., sugar maple, Acer saccharum) to over a thousand years (e.g., sacred lotus, Nelumbo nucifera). Successful germination requires the right conditions of light, water, and temperature and usually begins with imbibition of water and the subsequent release from dormancy. During its early growth stages and before it has become totally independent of the food stored in the seed or cotyledons, the new plant is called a seedling.

Two patterns of seed germination occur in angiosperms, depending on whether the cotyledons emerge from the seed: hypogeal (belowground germination) and epigeal (aboveground germination). In hypogeous germination, the hypocotyl remains short and the cotyledons do not emerge from the seed but rather force the radicle and epicotyl axis to elongate out of the seed coat. The seed, with the enclosed cotyledons, remains underground, and the epicotyl grows up through the soil. When the cotyledons contain seed-storage products, these products are transferred directly to the developing radicle and epicotyl (e.g., garden pea). When the endosperm or perisperm contains the storage products, the cotyledons penetrate the storage tissues and transfer the storage products to the developing radicle and epicotyl (e.g., garlic, Allium sativum).

In epigeous germination, the radicle emerges from the seed and the hypocotyl elongates, raising the cotyledons, epicotyl, and remains of the seed coat aboveground. The cotyledons may then expand and function photosynthetically as normal leaves (e.g., castor bean, Ricinus communis). When the cotyledons contain seed-storage products, they transfer them to the rest of the seedling and degenerate without becoming significantly photosynthetic (e.g., garden beans, Phaseolus). Eventually, the seedling becomes independent of the seed-storage products and grows into a mature plant capable of reproduction. Although the dispersal of seeds is essential in the reproduction and spread of angiosperm species, it is equally important for successful germination and seedling establishment to take place in an appropriate habitat.

Classification

Diagnostic Classification

The angiosperms are a well-characterized, sharply defined group. There is not a single living plant species whose status as an angiosperm or non-angiosperm is in doubt. Even the fossil record provides no forms that connect with any other group, although there are of course some fossils of individual plant parts that cannot be effectively classified.

Most typically, angiosperms are seed plants. This separates them from all other plants except the gymnosperms, of which the most familiar representatives are the conifers and cycads.

The ovules (forerunners of the seeds) of angiosperms are characteristically enclosed in an ovary, in contrast to those of gymnosperms, which are exposed to the air at the time of pollination and never enclosed in an ovary. Pollen of angiosperms is received by the stigma, a specialized structure that is usually elevated above the ovary on a more slender structure known as the style. Pollen grains germinate on the stigma, and the pollen tube must grow through the tissues of the style (if present) and the ovary to reach the ovule. The pollen grains of gymnosperms, in contrast, are received at an opening (the micropyle) atop the ovule.

The female gametophyte of angiosperms (called the embryo sac) is tiny and contains only a few (typically eight) nuclei; the cytoplasm associated more or less directly with these nuclei is not partitioned by cell walls. One of the several nuclei of the embryo sac serves as the egg in sexual reproduction, uniting with one of the two sperm nuclei delivered by the pollen tube. Two other nuclei of the embryo sac fuse with the second sperm nucleus from the pollen tube. This triple-fusion nucleus is characteristically the forerunner of a multicellular food-storage tissue in the seed, called the endosperm.

The process in which both nuclei from the pollen tube fuse is referred to as double fertilization. This is perhaps the most characteristic single feature of angiosperms and is not shared with any other group. Gymnosperms, in sharp contrast, have a multicellular female gametophyte that consists of many hundreds or even thousands of cells. Double fertilization does not take place in this case, and the female gametophyte develops into the food-storage tissue of the seed.

Furthermore, angiosperms have a more complex set of conducting tissues than do gymnosperms. The water-conducting tissue (xylem) ordinarily includes some long tubes called vessels. Only one small group of gymnosperms, the Gnetophyta, has vessels. The food-conducting tissue (phloem) of angiosperms characteristically has companion cells that bear a direct ontogenetic relationship to the sieve tubes through which the actual conduction takes place. The phloem of gymnosperms has less-specialized sieve cells and lacks companion cells.

Eudicots

One of the major changes in the understanding of the evolution of the angiosperms was the realization that the basic distinction among flowering plants is not between monocotyledon groups (monocots) and dicotyledon groups (dicots). Rather, plants thought of as being "typical dicots" have evolved from within another group that includes the more-basal dicots and the monocots together. This group of typical dicots is now known as the eudicots. Molecular-based evidence supports their being a single evolutionary lineage (monophyletic), and they are characterized by pollen that fundamentally has three furrows or pores (tricolpate), in contrast to the single pore or furrow of the monocot and basal dicot group (monosulcates).

Within the eudicots there is a large clade called the core eudicots, nearly all members of which show major differences in floral morphology from that of other flowering plants. In particular, the basic construction of the flower is much more stereotyped than in the basal eudicots, monocots, and basal dicots. Within nearly every order of the core eudicots, there are families with a basic "5 + 5 + (5 or 10) + (3 or 5)" floral construction. This refers to five sepals, five petals, one whorl of five stamens, often another whorl of five stamens, and finally a whorl of three or five carpels. The members of the whorls alternate with each other so that the petals are on radii midway between the sepal radii; the carpels in the centre of the flower are on the same radii as the sepals but are opposite to them. When core eudicots have only five stamens, as is common, these stamens usually are the stamens of the outer whorl—that is, they alternate with the petals and are opposite the sepals. Furthermore, the carpels at least are more or less fused, and there is often a well-developed nectary disc either surrounding the base of the ovary or, less frequently, borne immediately outside the stamens. The flowers are usually perfect and are radially symmetric. It is interesting that some families in most of the core eudicot orders, including Asterid orders such as Cornales and Ericales, have members with many stamens, but in nearly all cases these stamens develop in a different way than the numerous stamens in families such as Ranunculaceae (a basal eudicot) or Magnoliaceae (a basal angiosperm).

Core eudicots commonly show other features as well. Instead of the stamens having the pollen sac, or anther, attached at the base to a stalk, or filament, the two being more or less continuous, in core eudicots the filament is often attached at the back of the anther, and it narrows considerably just before it joins the anther. The pollen of the core eudicots commonly has three longitudinal depressions, or colpi, as does the pollen of the rest of the eudicots, but in the middle of each colpus there is a circular pore through which the pollen tube emerges; that is, the pollen is basically tricolporate. There are many deviations from this generalized structure, but keeping it in mind as a reference is helpful for understanding angiosperm evolution.

Within the core eudicots there are a number of major clades. These include the Asterids and Rosids, which are very species-rich, the former particularly so. The basic arrangement of the flower parts in these eudicot clades does not change, but the petals are commonly fused in the Asterids, forming a corolla tube. There are also chemicals common in the Asterids that are very rare in any other flowering plants.

Annotated Classification

The classification of flowering plants used here is a significant departure from the botanical classification system of the American botanist Arthur Cronquist (1981), which was based on similarities and differences in morphological, chemical, and anatomical characters. Since the early 1990s, studies of plant phylogeny have been transformed by molecular techniques, mainly those involving sequencing of segments of DNA from the chloroplasts and the nuclei of plant cells, as well as improved computational tools to analyze large amounts of data. Findings derived from the use of those techniques, which provided more robust and testable data on plant phylogeny, often conflicted with older, morphological-based schemes such as the Cronquist system. In 1998 a group of scientists who were participating in large-scale molecular analysis of flowering plants proposed a new overall classification system for the angiosperms. They called themselves the Angiosperm Phylogeny Group, and their new scheme became known as the APG system.

The APG system focused mainly on the level of families (with related families grouped into orders) because they are the groups around which most botanists organize their understanding of plant diversity. It need not be assumed, however, that different families or orders are equivalent in any evolutionary sense; rather, the APG organization signals a relative level in a hierarchy. Within any particular family, though, the system does presume, with some possible exceptions, that the genera included in it are all related and that the family itself is monophyletic (a lineage with all its members derived from a common ancestor); the same holds for the families included within a particular order. One of the main departures from the Cronquist system in the APG system is a less hierarchical arrangement of the higher-level groupings, which Cronquist divided into two classes: the monocotyledons (monocots), or Liliopsida, with five subclasses, and the dicotyledons (dicots), or Magnoliopsida, with six subclasses. The APG system does recognize some higher-level groupings but only at an informal level, such as eudicots, Rosids, and Asterids. It continues to recognize the monocots as a monophyletic group; however, they are now seen as having evolved from within a more-basal group of primitive dicotyledonous angiosperms. In contrast, Cronquist portrayed the monocots as being the sister group to all other dicotyledonous groups.

The APG system was not intended to be definitive, since some families were not included in the first large molecular analyses, and some of the relationships suggested were fairly tentative. Following the original APG publication, more families were added to the molecular analyses, allowing these families to be placed in orders, and other new studies called for adjustments in the circumscription of particular families and orders. Those changes were incorporated into an update in 2003 of the APG known as APG II, into another update in 2009 known as APG III, and into another revision in 2016 known as APG IV. The synopsis of flowering-plant classification presented here follows the APG IV system. It is important to recognize that modifications to the APG IV system continue as new data become available.

SEED DISPERSAL

Seed dispersal refers to the processes by which mature seeds disperse from the parent plant. Dispersal decreases competition with the parent and increases the likelihood of finding a suitable environment for growth. Sexual reproduction generally results in the production of fruits whose sole purpose is to enable the species to disperse and multiply. The part of the plant that acts in the dispersal is the diaspore (a term incorporating both fruit and seed). Although diaspore dispersal is the obvious end of reproduction, some plants rarely flower or set fruit and instead have evolved a very efficient system of vegetative reproduction by means of sucker shoots. Vegetative reproduction is very common in herbaceous plants that may spread by stolons , bulbils , or stem suckering. Still, most plants that reproduce vegetatively also reproduce sexually since this enables them to remain genetically variable and more adaptable to changes in the environment.

Wind Dispersal

The simplest form of seed dispersal is by wind and, not surprisingly, wind-dispersed fruits in temperate areas usually develop in breezy spring months. The same species that are wind-pollinated in temperate areas often bear wind-dispersed seed such as maple (Acer in Aceraceae), willow and poplar (Salix and Populus in Salicaceae), and ash (Fraxinus in Oleaceae). Typically the wind-dispersed seeds are developed quickly and dispersed in the same season. Wind-dispersed tree species are numerous in the warm, moist forests of the tropics—especially for tall trees in areas where there is a slight to prominent dry season. The height of the tree is important to enable the diaspore to catch the wind currents. The shape of wind-dispersed diaspores is often critical to their dispersal as well. Maple seeds have a samara and set up a whirling pattern as they fall, which may assist them in implantation. Poplar and willow seed are borne in a loose, cottony mass, which is extremely buoyant even in weak air currents. Although different morphological structures have evolved to disperse tree seed, the most common form of seed dispersal is wind-dispersed.

Typically, wind-dispersed species in tropical areas with seasonally dry periods lose their fruits late in the dry season or in the early rainy season that follows. This ensures adequate moisture for germinating seeds and adequate establishment before the next dry season. Wind-dispersed seeds have the distinct disadvantage of being at the peril of the elements. Most do not get carried very far away from the mother plant, and the population of insects that feed on the particular plant increases greatly at the time of flowering and can often destroy much of the crop. Some tropical species successfully avoid this by fruiting irregularly or even by what is known as mass fruiting, in which hundreds of individuals somehow manage to flower all at one time, literally swamping the predator population with more food than it can eat and thus preventing the insects from eating all the fruit.

The manner in which wind-dispersed diaspores are released is often critical to their dispersal. Because wet seeds do not float well on the air, most do not disperse except when the capsule is dry. Seeds are often contained within the capsule walls, and the valves of the capsule open increasingly further with only the uppermost seeds being capable of being blown free.

The same capsules may release seeds not from wind alone, but in part by mechanical motions and the inertia built up by movements of animals passing through a population. Each time a plant is bumped more seeds are cast away from it. This is a short distance but common type of seed dispersal in many prairie and forest edge plant populations.

Animal Dispersal

Animal-dispersed fruits are more common than wind-dispersed fruits and occur in species with a wider variety of life forms, including herbs, many vines, a modest number of tropical lianas, and shrubs as well as some trees. The morphology of animal-dispersed fruits varies depending on the organism doing the dispersal. The animals vary from those as small as ants to as large as horses or elephants.

Both birds and mammals are very effective dispersers. Birds are particularly effective dispersers since they can move the diaspore the farthest and the fastest. Diaspores dispersed by birds are usually colorful and lack any obvious scent (birds have keen vision but a poor sense of smell). Often the fruits feature contrasting colors so they are more easily seen. Frequently the outer covering of such fruit might be green or brown, but when the fruit opens the inner surface is bright red with a black seed. Often birds eat only the sweet portion of the diaspore and spit out the seed. If eaten, most seeds pass rather quickly through the bird's system and are ejected. Many times tiny colorful berries, such as those of Anthurium (Araceae), are initially quite sweet but quickly turn bitter after being eaten to encourage rejection. Anthurium also produces seeds with a sticky appendage that causes the seeds to stick to the bird's bill.

Mammal dispersers are common in both temperate and tropical areas. Mammal-dispersed diaspores are usually not particularly colorful but may be tasty and even have a distinct aroma when mature. (Mammals have only average sight compared to birds but typically have a good sense of smell.) Squirrels and rodents in temperate regions gather and hoard oak and hickory fruits while tropical agoutis "scatter hoard" fruits by burying them seemingly at random on the forest floor. Those fruits that are not found later in the late rainy season—when fruit is rare—are already planted and ready to grow. Monkeys in the tropics have a diverse diet with a broad array of fruits that are dispersed by them. They are amazingly adept at dispersal, but seemingly wasteful since they gather many fruits, eat part of them, and then discard the remainder along with the seeds. Some less common animal dispersers are horses, which are known to eat and disperse the seeds of calabash (Cresentia cujete in Bignoniaceae) in Central America.

In the case of epizoochorous fruits, animals are responsible for dispersing fruits without actually consuming them. These are diaspores that attach themselves to fur or clothing. Among the most effective types are beggar's ticks (Bidens in Asteraceae), tick-trefoil (Desmodium in Fabaceae), and Queen Anne's lace (Daucus carota in Apiaceae). These fruits are difficult to avoid and are difficult to remove, so they are usually picked off and discarded far from where they were first encountered.

Mechanical Dispersal

Mechanically dispersed seeds are common in both temperate and tropical areas. Many legumes (Fabaceae) have fruits that dry under torsion, and are suddenly released when the two halves of the fruits fall apart. In this instant the two halves of the valve twist laterally and sometimes also longitudinally, which causes the dry seeds along their length to be thrown for considerable distances. One of the most remarkable mechanically dispersed seeds is that of Hura crepitans (Euphorbiaceae), which is made up of a series of pie-shaped segments that burst open with such force that it sounds like a rifle shot. Its small flat seeds are carried for great distances.

Water Dispersal

Water dispersal is quite effective in estuarian populations of plants. The nature of water-dispersed fruits is important since a seed that lacks buoyancy would sink to the bottom near the mother plant and have to compete with it. A diaspore that was too buoyant would perhaps never sink at all and thus might never be implanted. Urospatha, a tropical aroid, has fruits with seeds that are embedded in a thick, buoyant, gelatinous mass, which allows them to float for a period and then sink into the water. The seeds of some tropical trees that occur along water courses are known to be consumed by fish. It is not yet known, however, whether the movement of the fish are important to the dispersal of the seeds.

Seed germination and the establishment of the young plant is, of course, the only true sign of reproductive success. Dispersal without establishment is to no avail. In every case the rate of germination is critical. Many dias-pores do not fall into the proper situation for germination. Often large numbers of seeds are killed by a wide variety of beetles or weevils that specialize on seeds. Different species have developed various methods of survival. Some, such as orchids, produce thousands of minute seeds per capsule, giving some a good chance of success. Other species use the opposite strategy of producing large and heavy fruits with a lot of stored food material to ensure survival after germination. Some species, such as the seeds of the Beilschmiedia in the Lauraceae, have an increased chance of survival by having the seeds begin the germination process while still on the trees, where they are less susceptible to attacks. The red mangrove Rhizophora mangle (Rhizophoraceae) goes even further by actually establishing a young plant on the tree that has a pointed base that actually implants in the soil when it falls.

Myrmecochory

Myrmecochory is seed dispersal by ants, an ecologically significant ant-plant interaction with worldwide distribution. Most myrmecochorous plants produce seeds with elaiosomes, a term encompassing various external appendages or "food bodies" rich in lipids, amino acid, or other nutrients that are attractive to ants. The seed with its attached elaiosome is collectively known as a diaspore. Seed dispersal by ants is typically accomplished when foraging workers carry diaspores back to the ant colony after which the elaiosome is removed or fed directly to ant larvae. Once the elaiosome is consumed the seed is usually discarded in underground middens or ejected from the nest. Although diaspores are seldom distributed far from the parent plant, myrmecochores also benefit from this predominantly mutualistic interaction through dispersal to favourable locations for germination as well as escape from seed predation.

Distribution and Diversity

Myrmecochory is exhibited by more than 3,000 plant species worldwide and is present in every major biome on all continents except Antarctica. Seed dispersal by ants is particularly common in the dry heath and sclerophyll woodlands of Australia (1,500 species) and the South African fynbos (1,000 species). Both regions have a Mediterranean climate and largely infertile soils (characterized by low phosphorus availability), two factors that are often cited to explain the distribution of myrmecochory. Myrmecochory is also present in mesic forests in temperate regions of the northern hemisphere (*i.e.* in Europe and in eastern North America) as well as in tropical forests and dry deserts, though to a lesser degree. Estimates for the true biodiversity of myrmecochorous plants range from 11,000 to as high as 23,000 species worldwide, or approximately 5% of all flowering plant species.

Ecology

Myrmecochory is usually classified as a mutualism but this is contingent on the degree to which participating species benefit from the interaction. It is likely that several different factors combine to create mutualistic conditions. Myrmecochorous plants may derive benefit from increased dispersal distance, directed dispersal to nutrient-enriched or protected microsites, and/or seed predator avoidance. Costs incurred by myrmecochorous plants include the energy required to provision diaspores, particularly when there is a disproportionate investment of growth-limiting mineral nutrients. For instance, some Australian *Acacia* species invest a significant portion of their yearly phosphorus uptake in producing diaspores. Diaspores must also be protected from outright predation by ants. This is typically accomplished by the production of a hard, smooth *testa*, or seed coat.

Few studies have examined the costs and benefits to ants participating in myrmecochory. Much remains to be understood about the selective advantages conferred upon myrmecochorous ants.

No single hypothesis explains the evolution and persistence of myrmecochory. Instead, it is likely that a combination of beneficial effects working at different spatio-temporal scales contribute to the viability of this predominantly mutualistic interaction. Three commonly cited advantages to myrmecochorous plants are increased dispersal distance, directed dispersal, and seed predator avoidance.

Dispersal Distance

Increasing dispersal distance from the parent plant is likely to reduce seed mortality resulting from density-dependent effects. Ants can transport seeds as far as 180 m but the average is less than 2 m, and values between 0.5 and 1.5 m are most common.

Perhaps due to the relatively limited distance that ants disperse seeds, many myrmecochores exhibit *diplochory*: a two-staged dispersal mechanism, often with ballistic projection as the initial mechanism, that can increase dispersal distance by as much as 50%. In some cases, ballistic dispersal distance regularly exceeds that of transport by ants. The dispersal distance achieved through myrmecochory is likely to provide an advantage proportionate to the spatial scale of density-dependent effects acting on individual plants. As such, the relatively modest distances ants transport seeds are likely to be more advantageous for myrmecochorous shrubs, forbs, and other plants of small stature.

Directed Dispersal

Myrmecochorous plants may benefit when ants disperse seeds to nutrient-rich or protected microsites that enhance germination and establishment of seedlings. Ants disperse seeds in fairly predictable ways, either by disposing of them in underground middens or by ejecting them from the nest. These patterns of ant dispersal are predictable enough to permit plants to manipulate animal behaviour and influence seed fate, effectively directing the dispersal of seeds to desirable sites. For example, myrmecochores can influence seed fate by producing rounder, smoother diaspores that inhibit ants from re-dispersing seeds after elaiosome removal. This increases the likelihood that seeds will remain underground instead of being ejected from the nest.

Nest chemistry is ideally suited for seed germination given that ant colonies are typically enriched with plant nutrients such as phosphorus and nitrate. This is likely to be advantageous in areas with infertile soils and less important in areas with more favourable soil chemistry, as in fertile forests. In fire-prone areas, depth of burial is an important factor for successful post-burn germination. This, in turn, is influenced by the nesting habits of the myrmecochorous ants. As such, the value of directed dispersal is largely context dependent.

Seed Predator Avoidance

Myrmecochorous plants escape or avoid seed predation by granivores when ants remove and sequester diaspores. This benefit is particularly pronounced in areas where myrmecochorous plants are subject to heavy seed predation, which may be common. In mesic forest habitats seed predators remove approximately 60% of all dispersed seeds within a few days and eventually remove all seeds not removed by ants. In addition to attracting ants, elaiosomes also appeal to granivores, and their presence can increase seed predation rates.

Nature of the Interaction

Myrmecochory is traditionally thought to be a diffuse or facultative mutualism with low specificity between myrmecochores and individual ant species. This assertion has been challenged in a study of Iberian myrmecochores demonstrating the disproportionate importance of specific ant species in dispersing seeds. Ant-plant interactions with a single species of myrmecochore were recorded for 37 species of ant but only 2 of these were found to disperse diaspores to any significant degree; the rest were seed predators or "cheaters" opportunistically feeding on elaiosomes *in situ* without dispersing seeds. Larger diaspores are hypothesized to increase the degree of specialization since ant mutualists need to be larger to successfully carry the diaspore back to the nest.

Ants, however, do not appear to form obligate relationships with myrmecochorous plants. Since no known ant species relies entirely on elaiosomes for their nutritional needs, ants remain generalist foragers even when entering into relationships with a more specialized myrmecochore.

As with many other facultative mutualisms, cheating is present on both sides of the interaction. Ants cheat by consuming elaiosomes without transporting seeds or through outright seed predation. Myrmecochorous plants can also cheat, either by producing diaspores with non-removable elaiosomes or by simulating the presence of a non-existent reward with chemical cues. Ants are sometimes capable of discriminating between cheaters and mutualists as shown by studies demonstrating preference for the diaspores of non-cheating myrmecochores. Cheating is also inhibited by ecological interactions external to the myrmecochorous interaction; simple models suggest that predation exerts a stabilizing influence on a mutualism such as myrmecochory.

Myrmecochory and Invasive Species

Myrmecochores are threatened by invasive species in some ecosystems. For instance, the Argentine ant is an aggressive invader capable of displacing native ant populations. Since Argentine ants do not disperse seeds, invasions may lead to a breakdown in the myrmecochory mutualism, inhibiting the dispersal ability of myrmecochores and causing long-term alterations in plant community dynamics. Invasive ant species can also maintain seed dispersal in their introduced range, as is the case with the red fire ant in the southeastern United States. Some invasive ants are also seed-disperses in their native range, such as the European fire ant, and can act as a high-quality disperses in their introduced range.

Myrmecochorous plants are also capable of invading ecosystems. These invaders may gain an advantage in areas where native ants disperse invasive seeds. Similarly, the spread of myrmecochorous invaders may be inhibited by limitations in the ranges of native ant populations.

Seed Dispersal Syndrome

A seed dispersal syndrome is a mutualistic plant-animal interaction. Seed dispersal syndromes are morphological characters of seeds correlated to particular seed dispersal agents. Dispersal is the event by which individuals move from the site of their parents to establish in a new area. A seed disperser is the vector by which a seed moves from its parent to the resting place where the individual will establish, for instance an animal. Similar to the term syndrome, a diaspore is a morphological functional unit of a seed for dispersal purposes.

Characteristics for seed dispersal syndromes are commonly fruit colour, mass, and persistence. These syndrome characteristics are often associated with the fruit that carries the seeds. Fruits are packages for seeds, composed of nutritious tissues to feed animals. However, fruit pulp is not commonly used as a seed dispersal syndrome because pulp nutritional value does not enhance seed dispersal success. Animals interact with these fruits because they are a common food source for them. Although, not all seed dispersal syndromes have fruits because not all seeds are dispersed by animals. Suitable biological and environmental conditions of dispersal syndromes are needed for seed dispersal and invasion success such as temperature and moisture.

Seed dispersal syndromes are parallel to pollination syndromes, which are defined as floral characteristics that attract organisms as pollinators. They are considered parallels because they are both plant-animal interactions, which increase the reproductive success of a plant. However, seed dispersal syndromes are more common in gymnosperms, while pollination syndromes are found in angiosperms. Seeds disperse to increase the reproductive success of the plant. The farther away a seed is from a parent, the better its chances of survival and germination. Therefore, a plant should select certain traits to increase dispersal by a vector (i.e. bird) to increase the reproductive success of the plant.

Types and Functions

Dispersal syndromes have been previously classified by: size, colour, weight, protection, flesh type, number of seeds, weight and start time of ripening. Syndromes are often associated with the type of dispersal and morphology. Also chemical composition can influence the disperser's fruit choice. The following are types of seed dispersal and their syndromes.

Anemochory

Anemochory is defined as seed dispersal by wind. Common dispersal syndromes of anemochory are wing structures and brown or dull coloured seeds without further rewards. Van der Pijl named seeds for anemochory flyers, rollers, or throwers to represent the seed dispersal syndromes and their behaviour. Flyers are typically categorized as dust diaspores, balloons, plumed or winged. Dust diaspores are small flat structures on seeds that appear to be the transition to wing diaspores, balloons are inflated seed characteristics and plumes are hairs or elongation seed characteristics. Wings have evolved to increase dispersal distance to promote gene flow. Anemochory is commonly found in open habitats, canopy trees, and dry season deciduous forests. Wind dispersers mature in the dry season for optimum high long-distance dispersal to increase success of germination.

Example of a syndrome of anemochory

Barochory

Barochory is seed dispersal by gravity alone in which a plant's seeds fall beneath the parent plant. These seeds commonly have heavy seed dispersal syndromes. However, heavy seeds may not be a form of seed dispersal syndrome, but a random seed characteristic that has no dispersal purpose. It has been thought that barochory does not develop a seed dispersal syndrome because it does not select for characters to enhance dispersal. It is questionable whether barochory is dispersal at all.

Hydrochory

Hydrochory is seed dispersal by water. Seeds can disperse by rain or ice or be submerged in water. Seeds dispersed by water need to have the ability to float and resist water damage. They often have hairs to assist with enlargement and floating. More features that cause floating are air space, lightweight tissues and corky tissues. Hydrochory syndromes are most common in aquatic plants.

Zoochory

Zoochory is the dispersal of seeds by animals and can be further divided into three classes.

1. Endozoochory is seed dispersal by animal ingestion and defecation of a seed. In a mutualistic behavior, the animal is rewarded with nutritious fruit while harmlessly dispersing the seed or seeds, thereby increasing their fitness and chances for survival.

2. Synzoochory is dispersal of diaspores by the mouthparts of animals.

3. Epizoochory is the accidental dispersal by animals. Differing characteristics of zoochory syndromes include coloured fruits, scented fruits, and different textures for different animals.

Endozoochory syndrome characteristics will develop based on palatability of the fruit by an organism. For example, mammals are attracted to scent of a seed and birds are attracted to colour. Endozoochory syndromes have evolved to be ingested by animals and later bypassed in a new environment so the seed can germinate. Synzoochory should possess hard skins to protect seeds from damage of mouthparts; for example, sharp beaks on animals such as birds or turtles. Epizoochory commonly has burrs or spines to transport seeds on the outside of animals. These syndromes are highly associated with animals that have fur, while burrs would be lacking on seeds that are dispersed by reptiles because of their smooth skin. It is believed that not all animals that interact with plant fruits are dispersers because some animals do not increase the successful dispersal of seeds but consume and destroy them. Therefore, some animals are dispersers and some are consumers.

Mammalochory

Mammalochory is specifically the seed dispersal by mammals. The dispersal syndromes for mammalochory include large fleshy fruit, green or dull coloured fruits, and husked or unhusked. The seeds tend to have more protection to prevent mechanical destruction. Mammals rely on smell more than vision for foraging, which causes the seeds they disperse to be more scented compared to bird-dispersed seeds. Animal-dispersed seeds ripen in rainy season when foraging activity is high, resulting in fleshy diaspores. Mammals consume fruits whole or in smaller pieces, which explains the larger seed syndromes. Mammalochory syndromes can increase the reproductive success of the plant compared to seed dispersal syndromes of a plant associated with barochory for example. An example of seed dispersal syndromes associated with mammals that increases reproductive success would be seed-consuming rodents that increase germination by burial of seeds.

Ornithochory

Ornithochory is seed dispersal by birds. Common syndrome characteristics include small fleshy fruits with bright colours and without husks. Ornithochory is common in temperate zones and oceanic islands because of absence of native mammals. Birds have heightened colour vision and swallow seeds and fruits whole, explaining the small and coloured characteristics of dispersal syndromes. Birds have a weak sense of smell, therefore ornithochory syndromes would specialize more in colour than scent, in comparison to mammalochory. Ornithochory can increase the reproductive success of a plant because a bird's digestive tract increases seed germination after it has been bypassed and dispersed by the bird.

Myrmecochory

Myrmecochory is seed dispersal by ants. Myrmecochory is considered an ant-plant mutualistic relationship. The common syndrome traits for myrmecochory are elaisomes, and are often hard and difficult to damage. Elaisomes are structures that attract ants because they are high in lipid content, providing important nutrients for the ant. Without ants, seed dispersal becomes barochory and dispersal success declines. It is debated if ants are good dispersers and if plants would select for ant dispersal. Ants do clearly interact with seeds, however ants cannot travel very long distances. Therefore, would a plant select for a bird over an ant when birds can disperse seeds much farther than ants, increasing a plant's reproductive success.

Problems in Seed Dispersal Syndromes

Many scientists are skeptical whether seed dispersal syndromes actually exist because their parallel, pollination syndromes, are often disputed in scientific literature. Seed dispersal syndromes do not have much disagreement among scientists. Whether this is due to lack of research or interest in seed dispersal syndromes, or that scientists agree with the idea of seed dispersal syndromes. It also may be that seed dispersal syndromes are harder to test because once seeds disperse they are difficult to collect and study. Jordano states that the evolution of fruit traits for seed dispersal success is only dependent on diameter. This is one scientist's perspective but does not appear to be the common consensus among scientists. Colour and olfaction are other common seed dispersal syndromes tested and discussed in scientific literature. One limitation to seed dispersal syndromes mentioned is the limited definitions of syndrome characteristics such as odour or texture. It is possible that there has not been enough research to test these characteristics or they do not play a role in seed dispersal syndromes.

The differences in seed dispersal syndromes appear to be weak, but do exist. There needs to be consideration for the possibility that these syndromes evolved not to benefit seed dispersal but possibility to combat other selective pressures. For example, syndromes may have developed to combat predation or environmental hazards. Predation could produce a secondary metabolite syndrome. Secondary metabolites are compounds that are not used for the primary function of a plant and are normally used as defense mechanisms

SEED TRAP

Seed traps are used in ecology and forestry to capture seeds falling from plants, allowing seed production and dispersal to be quantified. They come in several forms, including funnel traps, sticky traps (using materials such as fly paper), nets and pots exposed in the field.

Types of Trap

There are many options when using seed traps based on the specific need for the project. Seed traps can be made in different sizes, shapes, and of different material. Traditionally, seed traps are wooden frames with a screened bottom. Traps with metal frames have also been used. Additionally, funnel-shaped traps that are set above the soil or leveled at the ground, traps with screen or cloth bags, traps with water or soil to germinate them, plastic buckets, or traps with sticky surfaces have been used.

Sticky Traps

The substance on sticky traps must be nontoxic to the seeds and non-drying. Many sticky traps are petroleum-based. They are also cheaper (approximately $1.20–1.30 per use), lighter and less bulky than other traps. However, rain or dew may affect the adhesive. Heat and light intensities can also cause sticky traps to loose their adhesiveness.

Pollen Traps

Pollen traps are used to measure production and dispersal of pollen in plants. These traps are commonly made of glass slides with silicone oils or sticky tapes. They can be set up horizontally or as cylinder containers.

Fruit Traps

These traps can collect fruit production from various plants. These are also constructed on the ground, hanging or as basins or funnels.

Choosing a Trap

When choosing a trap, certain factors must be considered. The seed dispersal unit (where the seeds will fall), timing of dispersal, and density of seed fall (how much will be produced). Smaller traps may be appropriate for trees that produce more seeds, and larger traps may be appropriate for trees that produce less seeds to guarantee collection. It is difficult to compare different traps.

There are also some drawbacks to consider when choosing a trap. Wood traps may easily deteriorate if not constructed properly. Traps may also be hard to store or set up depending on size. Cloth and screens are easily torn in bad weather conditions and are also targets for animals. Screens may also harm many insects. Traps with water or soil must be maintained consistently. Extra seeds may clog the traps or seeds may be blown or washed out causing an over or underestimate of seed dispersal. Sticky traps are also clogged by other debris.

Maintenance

Traps must be checked weekly depending on the rate of seed dispersal. Seed identification must be accurate. This can be challenging because there is no comprehensive identification key, so focusing on only a few species at a time is important. You can also attempt to germinate the seeds. Screens may help prevent predation by animals and insects.

SEED DORMANCY

Seed dormancy is an evolutionary adaptation that prevents seeds from germinating during unsuitable ecological conditions that would typically lead to a low probability of seedling survival. Dormant seed do not germinate in a specified period of time under a combination of environmental factors that are normally conducive to the germination of non-dormant seeds.

An important function of seed dormancy is delayed germination, which allows dispersal and prevents simultaneous germination of all seeds. The staggering of germination safeguards some seeds and seedlings from suffering damage or death from short periods of bad weather or from transient herbivores; it also allows some seeds to germinate when competition from other plants for light and water might be less intense. Another form of delayed seed germination is seed quiescence, which is different from true seed dormancy and occurs when a seed fails to germinate because the external environmental conditions are too dry or warm or cold for germination. Many species of plants have seeds that delay germination for many months or years, and some seeds can remain in the soil seed bank for more than 50 years before germination. Some seeds have a very long viability period, and the oldest documented germinating seed was nearly 2000 years old based on radiocarbon dating.

True dormancy or innate dormancy is caused by conditions within the seed that prevent germination under normally ideal conditions. Often seed dormancy is divided into two major categories based on what part of the seed produces dormancy: exogenous and endogenous. There are three types of dormancy based on their mode of action: physical, physiological and morphological.

There have been a number of classification schemes developed to group different dormant seeds, but none have gained universal usage. Dormancy occurs because of a wide range of reasons that often overlap, producing conditions in which definitive categorization is not clear. Compounding this problem is that the same seed that is dormant for one reason at a given point may be dormant for another reason at a later point. Some seeds fluctuate from periods of dormancy to non dormancy, and despite the fact that a dormant seed appears to be static or inert, in reality they are still receiving and responding to environmental cues.

Exogenous Dormancy

Exogenous dormancy is caused by conditions outside the embryo and is often broken down into three subgroups:

Physical Dormancy

Dormancy caused by an impermeable seed coat is known as physical dormancy. Physical dormancy is the result of impermeable layer(s) that develops during maturation and drying of the seed or fruit. This impermeable layer prevents the seed from taking up water or gases. As a result, the seed is prevented from germinating until dormancy is broken. In natural systems, physical dormancy is broken by several factors including high temperatures, fluctuating temperatures, fire, freezing/thawing, drying or passage through the digestive tracts of animals. Physical dormancy is believed to have developed >100 mya.

Once physical dormancy is broken it cannot be reinstated (i.e. the seed is unable to enter secondary dormancy following unfavourable conditions unlike seeds with physiological dormancy mechanisms). Therefore, the timing of the mechanisms that breaks physical dormancy is critical and must be tuned to environmental cues. This maximises the chances for germination occurring in conditions where the plant will successfully germinate, establish and eventually reproduce.

Physical dormancy has been identified in the seeds of plants across 16 angiosperm families including:

- Anacardiaceae
- Asteraceae
- Bixaceae
- Cannaceae (monocot)
- Cistaceae
- Cochlospermaceae
- Convolvulaceae
- Cucurbitaceae
- Dipterocarpaceae
- Geraniaceae
- Legumeinosae
- Malvaceae
- Nelumbonaceae

- Rhamnaceae

- Sarcolaenaceae

- Sapindaceae

Physical dormancy has not been recorded in any gymnosperms. Generally, physical dormancy is the result of one or more palisade layers in the fruit or seed coat. These layers are lignified with malpighian cells tightly packed together and impregnated with water-repellent. In the families Anacardiaceae and Nelumbonaceae the seed coat is not well developed. Therefore, palisade layers in the fruit perform the functional role of preventing water uptake. While physical dormancy is a common feature, several species in these families do not have physical dormancy or produce non-dormant seeds.

Specialised structures, which function as a "water-gap", are associated with the impermeable layers of the seed to prevent the uptake of water. The water-gap is closed at seed maturity and is opened in response to the appropriate environmental signal. Breaking physical dormancy involves the disruption of these specialised structures within the seed, and acts as an environmental signal detector for germination. For example, legume (Fabaceae) seeds become permeable after the thin-walled cells of lens (water-gap structure). Following disrupted pulls apart to allow water entry into the seed. Other water gap structures include carpellary micropyle, bixoid chalazal plug, imbibition lid and the suberised 'stopper'.

In nature, the seed coats of physically dormant seeds are thought to become water permeable over time through repeated heating and cooling over many months-years in the soil seedbank. For example, the high and fluctuating temperatures during the dry season in northern Australia promote dormancy break in impermeable seeds of *Stylosanthes humilis* and *S.hamata* (Fabaceae).

Mechanical Dormancy

Mechanical dormancy when seed coats or other coverings are too hard to allow the embryo to expand during germination. In the past this mechanism of dormancy was ascribed to a number of species that have been found to have endogenous factors for their dormancy instead. These endogenous factors include low embryo growth potential.

Chemical Dormancy

Includes growth regulators etc., that are present in the coverings around the embryo. They may be leached out of the tissues by washing or soaking the seed, or deactivated by other means. Other chemicals that prevent germination are washed out of the seeds by rainwater or snow melt.

Endogenous Dormancy

Endogenous dormancy is caused by conditions within the embryo itself, and it is also often broken down into three subgroups: physiological dormancy, morphological dormancy and combined dormancy, each of these groups may also have subgroups.

Physiological Dormancy

Physiological dormancy prevents embryo growth and seed germination until chemical changes occur. Physiological dormancy is indicated when an increase in germination rate occurs after an application of gibberellic acid (GA3) or after Dry after-ripening or dry storage. It is also indicated when dormant seed embryos are excised and produce healthy seedlings: or when up to 3 months of cold (0–10 °C) or warm (=15 °C) stratification increases germination: or when dry after-ripening shortens the cold stratification period required. In some seeds physiological dormancy is indicated when scarification increases germination.

Physiological dormancy is broken when inhibiting chemicals are broken down or are no longer produced by the seed; often by a period of cool moist conditions, normally below (+4C) 39F, or in the case of many species in *Ranunculaceae* and a few others,(−5C) 24F. Abscisic acid is usually the growth inhibitor in seeds and its production can be affected by light. Some plants like Peony species have multiple types of physiological dormancy, one affects radicle (root) growth while the other affects plumule (shoot) growth.

- Drying: some plants including a number of grasses and those from seasonally arid regions need a period of drying before they will germinate, the seeds are released but need to have a lower moisture content before germination can begin. If the seeds remain moist after dispersal, germination can be delayed for many months or even years. Many herbaceous plants from temperate climate zones have physiological dormancy that disappears with drying of the seeds.

- Photodormancy or light sensitivity affects germination of some seeds: These photoblastic seeds need a period of darkness or light to germinate. In species with thin seed coats, light may be able to penetrate into the dormant embryo. The presence of light or the absence of light may trigger the germination process, inhibiting germination in some seeds buried too deeply or in others not buried in the soil.

- Thermodormancy is seed sensitivity to heat or cold: Some seeds including cocklebur and amaranth germinate only at high temperatures (30C or 86F). Many plants that have seeds that germinate in early to mid summer have thermodormancy and germinate only when the soil temperature is warm. Other seeds need cool soils to germinate, while others like celery are inhibited when soil temperatures are too warm. Often thermodormancy requirements disappear as the seed ages or dries.

Seeds are classified as having deep physiological dormancy under these conditions: applications of GA3 does not increase germination; or when excised embryos produce abnormal seedlings; or when seeds require more than 3 months of cold stratification to germinate.

Morphological Dormancy

In morphological dormancy, the embryo is underdeveloped or undifferentiated. Some seeds have fully differentiated embryos that need to grow more before seed germination, or the embryos are not differentiated into different tissues at the time of fruit ripening.

- Immature embryos: some plants release their seeds before the tissues of the embryos have fully differentiated, and the seeds ripen after they take in water while on the ground, germination can be delayed from a few weeks to a few months.

Combined Dormancy

These seeds have both morphological and physiological dormancy:

- Morpho-physiological or morphophysiological dormancy occurs when seeds with underdeveloped embryos, also have physiological components to dormancy. These seeds therefore require dormancy-breaking treatments as well as a period of time to develop fully grown embryos.

- Intermediate simple.

- Deep simple.

- Deep simple epicotyl.

- Deep simple double.

- Intermediate complex.

- Deep complex.

Combinational Dormancy

Combinational dormancy occurs in some seeds, where dormancy is caused by both exogenous (physical) and endogenous (physiological) conditions. some *Iris* species have both hard impermeable seeds coats and physiological dormancy.

Secondary Dormancy

Secondary dormancy occurs in some non-dormant and post dormant seeds that are exposed to conditions that are not favorable for germination, like high temperatures. It is caused by conditions that occur after the seed has been dispersed. The mechanisms of secondary dormancy are not yet fully understood but might involve the loss of sensitivity in receptors in the plasma membrane.

Not all seeds undergo a period of dormancy, many species of plants release their seeds late in the year when the soil temperature is too low for germination or when the environment is dry. If these seeds are collected and sown in an environment that is warm enough, and/or moist enough, they will germinate. Under natural conditions non dormant seeds released late in the growing season wait until spring when the soil temperature rises or in the case of seeds dispersed during dry periods until it rains and there is enough soil moisture.

Seeds that do not germinate because they have fleshy fruits that retard germination are quiescent, not dormant.

Many garden plants have seeds that will germinate readily as soon as they have water and are warm enough, though their wild ancestors had dormancy. These cultivated plants lack seed dormancy because of generations of selective pressure by plant breeders and gardeners that grew and kept plants that lacked dormancy.

Seeds of some mangroves are viviparous and begin to germinate while still attached to the parent; they produce a large, heavy root, which allows the seed to penetrate into the ground when it falls.

EMBRYOPHYTE

Embryophyta is a major grouping of plants, sometimes known as "land plants," that includes both the non-vascular bryophytes (mosses, hornworts, and liverworts) and the vascular land plants, which are those so familiar with their vascular system and true roots, leaves, and stems, such as the ferns, flowering plants, conifers, and ginkgos.

Embryophytes are characterized by an alternation of generations life cycle, apical cell growth, cuticle, antherida (male gametophyte organs), and archegonia (female gametophyte organs). Embryophytes are distinguished from the mostly aquatic algae, which do not develop embryos, nor have true roots, stems, or leaves, whereas the embryophytes do form embryos, and have differentiated stems and leaves, and in the case of the vascular plants, true roots.

The origin of embryophytes, as these multicellular plants arose and conquered the land, was a pivotal event in the history of life on earth. Without embryophytes, there would be no animals or humans surviving on the land portion of our planet. The provide food, habitat, energy, oxygen, protection, and numerous vital other functions for the world's creatures. Human beings also benefit from the aesthetic beauty, medicines, and innumerable products derived from these diverse plants.

Embryophytes are the most familiar group of plants. They include trees, flowers, ferns, mosses, and various other green land plants. All are complex multicellular eukaryotes with specialized reproductive organs. With very few exceptions, embryophytes obtain their energy through photosynthesis (that is, by absorbing light); and they synthesize their food from carbon dioxide.

Traditionally, plants were divided into the two groups of embryophytes (Embryobionta), which do develop embryos, and thallophytes (subkingdom Thallobionta), which do not develop embryos and which historically included both algae and fungi. However, fungi are no longer considered plants, and are placed in their own Kingdom. Note that most members of the thallophyte grouping do undergo alternation of generations, with two alternating generations, but all embryophytes undergo such alternation of generations.

Embryophyta may be distinguished from chlorophyll-using multicellular algae by having sterile tissue within the reproductive organs. Furthermore, embryophytes are primarily adapted for life on land, although some are aquatic (which some assume to be secondarily evolved). Accordingly, they are often called land plants or terrestrial plants.

The following are the synapomorphies of the embryophytes: a life cycle with an alternation of generations, apical cell growth (meristem-like growth organization), antheridia, archegonia, and a cuticle (outer covering used to control water loss on land).

On a microscopic level, embryophyte cells remain very similar to those of green algae. They are eukaryotic, with a cell wall composed of cellulose and plastids surrounded by two membranes. These usually take the form of chloroplasts, which conduct photosynthesis and store food in the form of starch, and characteristically are pigmented with chlorophylls a and b, generally giving them a bright green color. Embryophytes also generally have an enlarged central vacuole or tonoplast, which maintains cell turgor and keeps the plant rigid. They lack flagella and centrioles except in certain gametes.

Subgroups: Bryophytes and Vascular Plants

There are two major groupings within Embryophyta. Bryophytes, or nonvascular land plants, includes the mosses (division Bryophyta), hornworts (division Anthocerotophyta), and liverworts (division Marchantiophyta). Originally, the three groups were brought together as the three classes or three phyla within the division Bryophyta. However, since the three groups of bryophytes form a paraphyletic group, they now are placed in three separate divisions. They are grouped together as bryophytes because of their similarity as non-vascular land plants. Algae are also non-vascular, but are not "land plants." Note that bryophytes do require water to propagate, and thus live in water or moist habitats.

Like the vascular plants, bryophytes do have differentiated stems, and although these are generally only a few millimeters tall, they do provide mechanical support. They also have leaves, although these typically are one cell thick and lack veins. However, they lack true roots, with their filamentous rhizomes having a primary function of mechanical attachment rather than extracting soil nutrients.

Vascular plants comprise the other group of embryophytes. These are really the true land plants, distinguished by a vascular system, and also characterized by true stems, leaves, and roots. Vascular plants have specialized tissues for conducting water, with water transport occurring either in xylem or phloem. The xylem carries water and inorganic solutes upward toward the leaves from the roots, while phloem carries organic solutes throughout the plant. Vascular plants include the familiar ferns, clubmosses, horsetails, flowering plants (angiosperms), and conifers, and other gymnosperms. Scientific names for this group include Tracheophyta and Tracheobionta, but neither is very widely used.

Evolution and Classification

The higher-level classification of plants varies considerably. Some authors have restricted the kingdom Plantae to include only embryophytes, others have given them various names and ranks. The groups listed here are often considered divisions or phyla, but have also been treated as classes, and they are occasionally compressed into as few as two divisions. Some classifications, indeed, consider the term Embryophyta at the superphylum (superdivision) level, and include land plants and some Charophyceae in a subkingdom named Streptophyta.

Embryophytes are considered to have developed from complex green algae (Chlorophyta) during the Paleozoic era. The Charales, or stoneworts, appear to be the best living illustration of that developmental step. These alga-like plants undergo an alternation between haploid and diploid generations (respectively called gametophytes and sporophytes). In the first embryophytes, however, the sporophytes became very different in structure and function, remaining small and dependent on the parent for their entire brief life. Such plants are informally called bryophytes. They include three surviving groups:

- Bryophyta (mosses)

- Anthocerotophyta (hornworts)

- Marchantiophyta (liverworts)

All of the above bryophytes are relatively small and are usually confined to moist environments, relying on water to disperse their spores. Other plants, better adapted to terrestrial conditions, appeared during the Silurian period. During the Devonian period, they diversified and spread to many different land environments, becoming the vascular plants or tracheophytes.

Tracheophyta have vascular tissues or tracheids, which transport water throughout the body, and an outer layer or cuticle that resists drying out. In most vascular plants, the sporophyte is the dominant individual, and develops true leaves, stems, and roots, while the gametophyte remains very small.

Many vascular plants, however, still disperse using spores. They include two extant groups:

- Lycopodiophyta (clubmosses)

- Pteridophyta (ferns, whisk ferns, and horsetails)

Other groups, which first appeared towards the end of the Paleozoic era, reproduce using desiccation-resistant capsules called seeds. These groups are accordingly called spermatophytes or seed plants. In these forms, the gametophyte is completely reduced, taking the form of single-celled pollen and ova, while the sporophyte begins its life enclosed within the seed. Some seed plants may even survive in extremely arid conditions, unlike their more water-bound precursors. The seed plants include the following extant groups:

- Cycadophyta (cycads)

- Ginkgophyta (ginkgo)

- Pinophyta (conifers)

- Gnetophyta (gnetae)

- Magnoliophyta (flowering plants)

The first four groups are referred to as gymnosperms, since the embryonic sporophyte is not enclosed until after pollination. In contrast, among the flowering plants or angiosperms, the pollen has to grow a tube to penetrate the seed coat. Angiosperms were the last major group of plants to appear, developing from gymnosperms during the Jurassic period, and then spreading rapidly during the Cretaceous. They are the predominant group of plants in most terrestrial biomes today.

References

- Walters, Dirk R Walters Bonnie By (1996). Vascular plant taxonomy. Dubuque, Iowa: Kendall/Hunt Pub. Co. p. 124. ISBN 978-0-7872-2108-9

- Spermatophytes, plants, argomento, plants-knowledge: eniscuola.net, Retrieved 21 June, 2019

- Stuble, K. L.; Kirkman, L. K.; Carroll, C. R. (2010). "Are red imported fire ants facilitators of native seed dispersal?". Biological Invasions. 12 (6): 1661–1669. doi:10.1007/s10530-009-9579-0.

- Angiosperm, plant: britannica.com, Retrieved 11 April, 2019

- Hollander, J.L. & Vander Wall, S.B. (2009). Dispersal syndromes in North American Ephedra. International Journal of Plant Sciences

- Daniel Simberloff; Marcel Rejmánek, eds. (2010). Encyclopedia of biological invasions. Berkeley: University of California Press. p. 730. ISBN 9780520948433

- Seed-dispersal, science, news-wires-white-papers-and-books: encyclopedia.com, Retrieved 17 May, 2019

- William J. Sutherland (3 August 2006). Ecological Census Techniques: A Handbook. Cambridge University Press. pp. 202–. ISBN 978-0-521-84462-8. Retrieved 22 April, 2012

- Embryophyte, entry: newworldencyclopedia.org, Retrieved 08 July, 2019

Seed Germination

The process through which a plant grows from a seed is known as germination. The germination of plant can take place underground or above the ground. Underground germination is known as hypogeal germination and above ground germination is known as epigeal germination. All these diverse aspects of germination have been carefully analyzed in this chapter.

Germination is the sprouting of a seed, spore, or other reproductive body, usually after a period of dormancy. The absorption of water, the passage of time, chilling, warming, oxygen availability, and light exposure may all operate in initiating the process.

In the process of seed germination, water is absorbed by the embryo, which results in the rehydration and expansion of the cells. Shortly after the beginning of water uptake, or imbibition, the rate of respiration increases, and various metabolic processes, suspended or much reduced during dormancy, resume. These events are associated with structural changes in the organelles (membranous bodies concerned with metabolism), in the cells of the embryo.

Germination sometimes occurs early in the development process; the mangrove (Rhizophora) embryo develops within the ovule, pushing out a swollen rudimentary root through the still-attached flower. In peas and corn (maize) the cotyledons (seed leaves) remain underground (e.g., hypogeal germination), while in other species (beans, sunflowers, etc.) the hypocotyl (embryonic stem) grows several inches above the ground, carrying the cotyledons into the light, in which they become green and often leaflike (e.g., epigeal germination).

Seedling Emergence

Active growth in the embryo, other than swelling resulting from imbibition, usually begins with the emergence of the primary root, known as the radicle, from the seed, although in some species (e.g., the coconut) the shoot, or plumule, emerges first. Early growth is dependent mainly upon cell expansion, but within a short time cell division begins in the radicle and young shoot, and thereafter growth and further organ formation (organogenesis) are based upon the usual combination of increase in cell number and enlargement of individual cells.

Until it becomes nutritionally self-supporting, the seedling depends upon reserves provided by the parent sporophyte. In angiosperms these reserves are found in the endosperm, in residual tissues of the ovule, or in the body of the embryo, usually in the cotyledons. In gymnosperms food materials are contained mainly in the female gametophyte. Since reserve materials are partly in insoluble form—as starch grains, protein granules, lipid droplets, and the like—much of the early metabolism of the seedling is concerned with mobilizing these materials and delivering, or translocating, the products to active areas. Reserves outside the embryo are digested by enzymes secreted by the embryo and, in some instances, also by special cells of the endosperm.

In some seeds (e.g., castor beans) absorption of nutrients from reserves is through the cotyledons, which later expand in the light to become the first organs active in photosynthesis. When the reserves are stored in the cotyledons themselves, these organs may shrink after germination and die or develop chlorophyll and become photosynthetic.

Environmental factors play an important part not only in determining the orientation of the seedling during its establishment as a rooted plant but also in controlling some aspects of its development. The response of the seedling to gravity is important. The radicle, which normally grows downward into the soil, is said to be positively geotropic. The young shoot, or plumule, is said to be negatively geotropic because it moves away from the soil; it rises by the extension of either the hypocotyl, the region between the radicle and the cotyledons, or the epicotyl, the segment above the level of the cotyledons. If the hypocotyl is extended, the cotyledons are carried out of the soil. If the epicotyl elongates, the cotyledons remain in the soil.

Light affects both the orientation of the seedling and its form. When a seed germinates below the soil surface, the plumule may emerge bent over, thus protecting its delicate tip, only to straighten out when exposed to light (the curvature is retained if the shoot emerges into darkness). Correspondingly, the young leaves of the plumule in such plants as the bean do not expand and become green except after exposure to light. These adaptative responses are known to be governed by reactions in which the light-sensitive pigment phytochrome plays a part. In most seedlings, the shoot shows a strong attraction to light, or a positive phototropism, which is most evident when the source of light is from one direction. Combined with the response to gravity, this positive phototropism maximizes the likelihood that the aerial parts of the plant will reach the environment most favourable for photosynthesis.

PHYSIOLOGY OF SEED GERMINATION

All the viable seeds which have overcome dormancy (if any) either naturally or artificially will readily germinate under suitable environmental conditions necessary for seed germination i.e., water, O_2, temperature and in some cases light. Such seeds which just wait for suitable environmental conditions to germinate are said to be 'quiescent'.

In most cases these seeds germinate if placed on moist substrate. The process of seed germination starts with the imbibition of water by seed coats and emergence of growing root tip of embryo. It ends when the embryo has developed into a seedling which is out of bounds of seed coats and has its own photosynthetic system. Before describing the physiological and biochemical changes accompanying the seed germination, it is better to evaluate the physiological state of seed immediately before germination.

Physiological Condition of Quiescent Seed

Before germination seed is a dry structure with various metabolic activities reduced to a minimum. It has dry, comparatively hard seed coat consisting of usually non-living cells. This seed coat and the cells of endosperm when present form a barrier to the outward growth of the embryo.

Moreover, in certain seeds the seed coat is impermeable to water and O_2 and acts as barrier between embryo and these substances. Therefore, the cells of the seed coat and endosperm (when present) must become permeable to water and oxygen, also these must become penetrable to the growing root tip of the embryo if germination is to occur. Most part of the embryo (excluding cotyledons) consists of potentially meristematic cells. But still these cells do not divide and enlarge and have minimum respiration rate in dry seed.

It is chiefly due to the following reasons:

- In absence of sufficient amount of water, these cells are unable to maintain turgor so that their growth is checked.

- These cells do not have sufficient amount of soluble respirable food. The reserve food stored in cotyledons or endosperm is in insoluble form and is not available to these cells.

- Aerobic respiration in embryo cells is at its minimum. It is because seed coat acts as barrier to O_2. Oxygen uptake in dry seeds is reduced to about 0.05 μl/g tissue/hr.

- Seed coat may contain inhibitors which check growth of these cells.

- The concentration of hydrolytic enzymes is low in dry seeds.

- Hard seed coat forms a physical barrier to the growth of embryo.

All these above conditions are admirably overcome if seeds are placed under suitable conditions essential for germination and in most cases seed germination begins just by placing the dry seeds on a moist substrate.

Physiological, biochemical and other changes accompanying seed germination:

Water Uptake

Seed germination, as mentioned earlier, starts with the imbibition of water by dry seed coat which is purely a physical process. Various hydrophilic groups such as $-NH_2$, $-OH$, $-COOH$ etc., of proteins, polymeric carbohydrates etc., found in the seed coat attract dipolar water molecules and form hydrated shells around them resulting in the swelling of these substances. This water uptake by swelling is followed by intensive water uptake associated with germination. Due to imbibition of water the seed coats become (i) more permeable to O_2 and water and (ii) less resistant to outward growth of the embryo.

Respiration

The uptake is accompanied by rapid increase in respiration rate of embryo. Initially there may be anaerobic respiration but it is soon replaced by aerobic one due to availability of O_2. As compared to dry seeds, the uptake of O_2 in germination seeds may rise in case of cereals from 0.05 µ 1/g tissue/hr to 100 µ 1/g tissue/hr within very short period after germination when water content has reached about 40%. Sucrose is probably the respiratory substrate at this stage which is provided by endosperm.

Mobilization of Reserve Materials

As germination progresses there is mobilization of reserve materials to provide:

- Building blocks for the development of embryo,

- Energy for the biosynthetic processes, and

- Nucleic acids for control of protein synthesis and overall embryonic development.

Nucleic Acids

In monocots during the imbibition stage of seed germination there is rapid decrease of DNA and RNA content in the endosperm with a simultaneous increase in the embryonic axis probably due to their transportation as such. Appreciable amount of RNA appears in the aleurone layer after about 16 hours which is probably due to its de novo synthesis. Higher concentration of RNA (and also protein) in the embryonic axis precedes cell division. Due to more cell divisions the DNA content is increased.

Carbohydrates

Insoluble carbohydrates like starch are the important reserve food of cereals in the endosperm. During germination starch is hydrolysed first into maltose in the presence of α-amylase and β-amylase and then the maltose is converted into glucose by maltase. The glucose is absorbed by the scutellum, converted into soluble sucrose and transported to growing embryonic axis.

During germination the embryonic axis secretes gibberellic acid into the aleurone layer which causes de novo synthesis of a-amylase. This enzyme is not found in un-germinated seeds. Removal of gibberellic acid during this period results in rapid fall of α-amylase synthesis. The latter can be restored by supplying external gibberellic acid. The gibberellic acid induced synthesis of α-amylase is countered by abscisic acid. In contrast to α-amylase, β-amylase is already present in the seed in inactive form which gets activated during germination. The activity of the enzyme maltase is also regulated by gibberellic acid.

Lipids

Many plants like castor bean, peanut etc. store large amount of neutral lipids or fats as reserve food in their seeds. During germination the mobilization of these fats is brought about by hydrolysis of fats to fatty acids and glycerol by lipases and P-oxidation of fatty acids to acetyl-CoA. The activity of lipases is greatly stimulated by imbibition and in some cases there may even be de novo synthesis of these enzymes which is probably triggered by gibberellic acid.

Some of the acetyl-CoA is converted into sucrose via the glyoxylate cycle and is transported to the growing embryonic axis. Synthesis of the two key enzymes of glyoxylate cycle (i.e., isocitratase and malate synthetase) takes place de novo during the early stages of germination. Their concentration is increased during the stage when fats are being actively converted into sucrose. They disappear when all the stored fat has been consumed and the seedling has developed the photosynthetic system.

Proteins

Some plants store proteins as reserve food in their seeds in the form of aleurone grains. Mobilization of these proteins involves their hydrolytic cleavage into amino acids by peptidases. These enzymes, in part, are synthesized de novo as the germination starts in the same way as a-amylase is synthesized in cereals. The amino acids may either provide energy by oxidation after deamination or may be utilised in the synthesis of new proteins.

During seed germination there is active synthesis of enzymes and other proteins and also the formation of different RNA species for their synthesis. For the synthesis of these proteins polysomes are involved. It is not definite whether these are already present in the seed or are synthesized during germination. Moreover, it is also not certain whether polysomes are synthesized de novo or from pre-existing m-RNA.

Inorganic Materials

A number of inorganic materials such as phosphate, calcium, magnesium and potassium are also stored in seeds in the form of phytin. These materials which may activate a number of important enzymes are liberated during germination due to the activity of various phosphatases including phytase.

Emergence of Seedling out of the Seed Coat

All these changes described above gradually result in splitting of seed coat and emergence of the growing seedling. First, the radicle comes out and grows downward, then plumule comes out and grows upward. Due to continued growth of this seedling, the latter comes out of the soil, exposed to light and develops its own photosynthetic apparatus.

The splitting of seed coat may take place either:

- By imbibitional pressure.

- By internal pressure created by the growing primary root.

- By hydrolytic enzymes which act on cell wall contents of seed coat and digest it e.g., cellulase, pectinase etc. Sometimes the seed coat may be extensively rotted by the activity of micro-organisms in the soil.

EPIGEAL GERMINATION

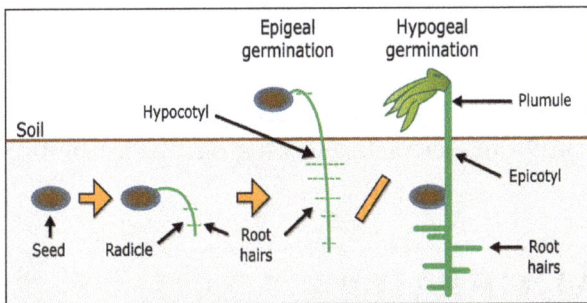

Epigeal vs. hypogeal germination

Epigeal germination is a botanical term indicating that the germination of a plant takes place above the ground. An example of a plant with epigeal germination is the common bean (*Phaseolus vulgaris*). The opposite of epigeal is hypogeal (underground germination). Epigeal is also not the same as hypogeal germination; both epigeal and hypogeal plants will grow differently.

Epigeal germination implies that the cotyledons are pushed above ground. The hypocotyl elongates while the epicotyl remains the same in length. In this way, the hypocotyl pushes the cotyledon upward.

Normally, the cotyledon itself contains very little nutrients in plants that show this kind of germination. Instead, the first leaflets are already folded up inside it, and photosynthesis starts to take place in it rather quickly.

Because the cotyledon is positioned above the ground it is much more vulnerable to damage like night-frost or grazing. The evolutionary strategy is that the plant produces a large number of seeds, of which statistically a number survive.

Plants that show epigeal germination need external nutrients rather quickly in order to develop, so they are more frequent on nutrient-rich soils. The plants also need relatively much sunlight for photosynthesis to take place. Therefore they can be found more often in the field, at the border of forests, or as pioneer species.

Plants that show epigeal germination grow relatively fast, especially in the first phase when the leaflets unfold. Because of this, they occur frequently in areas that experience regular flooding, for example at the river borders in the Amazon region. The fast germination enables the plant to develop before the next flooding takes place. After the faster first phase, the plant develops more slowly than plants that show hypogeal germination.

It is possible that within the same genus one species shows epigeal germination while another species shows hypogeal germination. Some genera in which this happens are:

- Phaseolus: The common bean (Phaseolus vulgaris) shows epigeal germination, whereas the runner bean (Phaseolus coccineus) shows hypogeal germination,

- Lilium,

- Araucaria: Species in the section Eutacta show epigeal germination, whereas species in the section Araucaria show hypogeal germination.

HYPOGEAL GERMINATION

Hypogeal germination is a botanical term indicating that the germination of a plant takes place below the ground. An example of a plant with hypogeal germination is the pea (Pisum sativum). The opposite of hypogeal is epigeal (above-ground germination).

Hypogeal germination implies that the cotyledons stay below the ground. The epicotyl (part of the stem above the cotyledon) grows, while the hypocotyl (part of the stem below the cotyledon) remains the same in length. In this way, the epicotyl pushes the plumule above the ground. Normally, the cotyledon is fleshy, and contains many nutrients that are used for germination. No photosynthesis takes place within the cotyledon.

Because the cotyledon stays below the ground, it is much less vulnerable to for example night-frost or grazing. The evolutionary strategy is that the plant produces a relatively low number of seeds, but each seed has a bigger chance of surviving.

Plants that show hypogeal germination need relatively little in the way of external nutrients to grow, therefore they are more frequent on nutrient-poor soils. The plants also need less sunlight, so they can be found more often in the middle of forests, where there is much competition to reach the sunlight.

Plants that show hypogeal germination grow relatively slowly, especially in the first phase. In areas that are regularly flooded, they need more time between floodings to develop. On the other hand, they are more resistant when a flooding takes place. After the slower first phase, the plant develops faster than plants that show epigeal germination. It is possible that within the same genus one species shows hypogeal germination while another species shows epigeal germination. Some genera in which this happens are:

- Phaseolus: the runner bean (Phaseolus coccineus) shows hypogeal germination, whereas the common bean (Phaseolus vulgaris) shows epigeal germination,

- Lilium,

- Araucaria: species in the section *Araucaria* show hypogeal germination, whereas species in the section *Eutacta* show epigeal germination.

FACTORS AFFECTING SEED GERMINATION

Some of the important factors are: (1) External factors such as water, oxygen and suitable temperature. (2) Internal factors such as seed dormancy due to internal conditions and its release.

External Factors

Water

A dormant seed is generally dehydrated and contains hardly 6-15% water in its living cells. The active cells, however, require about 75-95% of water for carrying out their metabolism. Therefore, the dormant seeds must absorb external water to become active and show germination. Besides providing the necessary hydration for the vital activities of protoplasm, water softens the seed coats, causes their rupturing, increases permeability of seeds, and converts the insoluble food into soluble form for its translocation to the embryo. Water also brings in the dissolved oxygen for use by the growing embryo.

Oxygen

Oxygen is necessary for respiration which releases the energy needed for growth. Germinating seeds respire very actively and need sufficient oxygen. The germinating seeds obtain this oxygen from the air contained in the soil. It is for this reason that most seeds sown deeper in the soil or in water-logged soils (i.e. oxygen deficient) often fail to germinate due to insufficient oxygen. Ploughing and hoeing aerate the soil and facilitate good germination.

Suitable Temperature

Moderate warmth is necessary for the vital activities of protoplasm, and, therefore, for seed germination. Though germination can take place over a wide range of temperature (5-40 °C), the optimum for most of the crop plants is around 25-30 °C. The germination in most cases tops at 0 °C and 45 °C.

Internal Factors

Seed Dormancy due to Internal Conditions and its Release

In some plants the embryo is not fully mature at the time of seed shedding. Such seeds do not germinate till the embryo attains maturity. The freshly shed seed in certain plants may not have sufficient amounts of growth hormones required for the growth of embryo. These seeds require some interval of time during which the hormones get synthesized.

The seeds of almost all the plants remain viable or living for a specific period of time. This viability period ranges from a few weeks to many years. Seeds of Lotus have the maximum viability period of 1000 years. Seeds germinate before the ending of their viability periods.

In many plants, the freshly shed seeds become dormant due to various reasons like the presence of hard, tough and impermeable seed coats, presence of growth inhibitors and the deficiency of sufficient amounts of food, minerals and enzymes, etc.

References

- Germination, science: britannica.com, Retrieved 20 May, 2019

- Franceschini, M. (2004) "An unusual case of epigeal cryptocotylar germination in Rollinia salicifolia (Annonaceae)" Botanical Journal of the Linnean Society vol. 146 no. 1

- Verbalization, plants, physiology-of-seed-germination-23596: biologydiscussion.com, Retrieved 30 June, 2019

- Germination, seed, factors-affecting-seed-germination-external-and-internal-factors-15758: biologydiscussion.com, Retrieved 11 March, 2019

Sowing Techniques and Methods

The process of planting seeds by scattering them on or inside the earth is known as sowing. Some of the techniques which are used to sow seeds are dibbling, aerial seeding, precision seeding and drilling. All the diverse methods and techniques related to sowing have been carefully analyzed in this chapter.

DIBBLING

Dibbling is a mode of sowing grain, es pecially wheat, much practised in some parts of England. It is found to answer the best on the clover leys of the lighter descriptions of land. It is performed by a man walking backwards with an iron dibble into each hand, with which he makes the holes, on the furrow slice, into which the seed is dropped by child ren, who place one or two seeds into each hole. By this mode there is a very considerable saving of seed, the quantity employed of wheat being usually from three to five pecks. The wheat plant obtains a more solid soil, and con siderable additional employment is afforded to the labourer and his family. It is, however, a rather tedious process, and is not adapted to the stiffer descriptions of soil, for on these the dibble forms little cups, in which the rain is apt to lodge to the destruction of the seed grain. A good dibbler with three active at tendants will plant about half an acre per day. The expense for labour is commonly about 7s. to 9s. per acre for wheat.

Advantages

For such method of seeding, the entire field need not be prepared for the seedbed but only the seeding zone. Moreover:

- It facilitates the practice of conservative tillage and reduces the chances of soil erosion.

- It requires less seeds and it gives rapid and uniform germination with good seedling vigour.

- Intercultural practices like weeding, earthing up and care of individual plants can be facilitated.

- When proper and uniform spacing is maintained, it becomes very easy to calculate the plant population and thereby expected yield.

Disadvantages

Uniform germination is not possible if all seeds are not placed at uniform depth. Besides, dibbling is a more laborious, time consuming and expensive process compared with broadcasting.

Field condition for dibbling: Seeds are sown in dry or semi-dry soil conditions and manures and fertilizers including pesticides and soil conditioners may be applied simultaneously. This method is suitable for planting maize, cotton, castor, potato, groundnut, pig soybean, cow-pea, soybean, sunflower, sugarcane, sweet potato, onion, garlic, turmeric, ginger, gourds, napier and guinea grass.

BROADCAST SEEDING

In agriculture, gardening, and forestry, broadcast seeding is a method of seeding that involves scattering seed, by hand or mechanically, over a relatively large area. This is in contrast to:

- Precision seeding, where seed is placed at a precise spacing and depth;

- Hydroseeding, where a slurry of seed, mulch and water is sprayed over prepared ground in a uniform layer.

Broadcast seeding is of particular use in establishing dense plant spacing, as for cover crops and lawns. In comparison to traditional drill planting, broadcast seeding will require 10–20% more seed. It's simpler, faster, and easier than traditional row sowing. Broadcast seeding works best for plants that do not require singular spacing or that are more easily thinned later. After broadcasting, seed is often lightly buried with some type of raking action, often done using vertical tillage tools. Utilizing these tools increases the success rate of germination by increasing seed-to-soil contact.

Seeds sown in this manner are distributed unevenly, which may result in overcrowding. This method may not ensure that all seeds are sown at the correct depth. Incorrect depth, if too deep, would result in germination that would not allow the young plant to break the surface of the soil and prevent sprouting. If they are not sown evenly then there would be a lack of various nutrients from sunlight, oxygen etc in many crops or plants.

In addition, it is worth noting that not all seeds are good candidates for broadcast seeding. Often, only smaller seeds will sprout and continue to grow successfully when planted by way of broadcasting. In general, the larger the seed, the deeper it can be planted.

AERIAL SEEDING

Aerial seeding is a technique of sowing seeds by spraying them through aerial mechanical means such as a drone, plane or helicopter.

Aerial seeding is considered a broadcast method of seeding. It is often used to spread different grasses and legumes to large areas of land that are in need of vegetative cover after fires. Large wildfires can destroy large areas of plant life resulting in erosion hazards. Aerial seeding may quickly and effectively reduce erosion hazards and suppress growth of invasive plant species. Aerial seeding is an alternative to other seeding methods where terrain is extremely rocky or at high elevations or otherwise inaccessible.

Helicopter Aerial Seeding

Plane Aerial Seeding

Aerial seeding is also often used to plant cover crops. Below is a small list of plants that are often seeded by this method.

Perennial Rye, Sudan grass	Soybeans, Buckwheat	Hairy Vetch, Corn
Cereal Rye, Winter Wheat	Oats, Mammoth or medium Red Cover	Sweet Clover, Berseem Clover
Crimson Clover, Perennial Rye Timothy	Perennial Rye Red Fescue, Perennial Rye Bluegrass	Perennial Rye Red Top, Red Clover Timothy

Major Advantage

The major advantage of aerial seeding is the efficient coverage of a large area in the least amount of time. Aerial seeding facilitates seeding in areas that otherwise would be impossible to seed with traditional methods, such as land that is too hard to reach by non aircraft or ground conditions being far too wet. Aerial seeding may be used when existing crops are already planted. This is important when living in an area where there is a small window between harvesting the crop and the end of the growing season, because seeding cover crops after harvest can cause poor stand establishment due to cold temperatures or moisture.

Soil Conditions

Soil moisture plays a large role in the success of aerial seeding. Adequate soil moisture for germination and establishment of seed requires that the top 0.5 -1 inch be moist. These conditions should be at the time of the seeding or within 10 days of the seeding. If the required soil moisture is not present at these times, the seeds may become the target of predation by insects and other animals. Along with soil moisture, surface conditions also play a key role in the success of aerial seeding and establishment of seed.

The best soil surface conditions are those that are moist and friable. A loose and rough soil surface with cracks or residue cover is also very conducive to seed germination. These conditions allow for the seed to make the best contact with moist soil while adequately allowing the seed to settle into the ground. Another important factor besides soil surface conditions are that of timing of aerial seeding and seeding rates.

Timing and Seeding Rates

When aerial seeding a cover crop one must seed them at least 7 to 10 days before drilled cover crops. The reason for this is because the aerial seeding method is slower than that of the drilled method. Seeding rates for most plants should be 25% to 50% higher with aerial seeding when compared to other more conventional methods like drilling. These higher seeding rates are needed to establish same yields as other methods. This is mostly because with aerial seeding the seed can be often on the soil surface longer making the seed more susceptible to predation by birds, insects and other animals.

Helicopter vs. Plane

There has been much debate with regards to which type of aircraft is better for aerial seeding. There is some evidence that shows helicopters may be best at the job when seeding in fields that are already established. This is because the wind from the blades of the helicopter causes the canopy of the already established crop to shake and open. This allows for more seed to reach the soil below. Another advantage for the helicopter is they are more maneuverable and can handle irregular shape fields while a plane has a harder time with such fields. The real advantage of the plane is it is faster than a helicopter and has the ability to carry a much heavier load. This allows the plane to finish the job much faster which equals less money spent in the air.

PRECISION SEEDING

In agriculture, precision seeding is a method of seeding that involves placing seed at a precise spacing and depth. This is in contrast to broadcast seeding, where seed is scattered over an area.

Although precise hand placement would qualify, precision seeding usually refers to a mechanical process. A wide range of hand-push and powered precision seeders are available for small- to large-scale jobs. Using a variety of actions, they all open the soil, place the seed, then cover it, to create rows. The depth and spacing vary depending on the type of crop and the desired plant density.

In commercial production, precision seeding is an alternative to placing larger quantities of seed in a row, by dribbling seed or setting several seeds in each position. Depending on the device, precision seeders may place only one, or a very few seeds per position. This is an advantage, in that it saves seed and it avoids crowding, or the need for thinning, allowing plants the space to grow efficiently. On the downside, by placing fewer seeds, a very high germination rate is required to make full use of the seeded area.

DRILLING

Drilling seeds is when you use a drill machine, which sows seeds evenly and at a particular spacing. Dibbling is when you use a frame with pegs at certain intervals, which you press into the ground manually, making holes in the soil at particulate spacing and intervals, and placing the seeds into the holes manually.

Seed Drill

A seed drill is a device that sows the seeds for crops by positioning them in the soil and burying them to a specific depth. This ensures that seeds will be distributed evenly.

The seed drill sows the seeds at the proper seeding rate and depth, ensuring that the seeds are covered by soil. This saves them from being eaten by birds and animals, or being dried up due to exposure to sun. With seed drill machines, seeds are distributed in rows, however the distance between seeds along the row cannot be adjusted by the user as in the case of vacuum precision planters. The distance between rows is typically set by the manufacturer. This allows plants to get sufficient sunlight, nutrients, and water from the soil. Before the introduction of the seed drill, most seeds were planted by hand broadcasting, an imprecise and wasteful process with a poor distribution of seeds and low productivity. Use of a seed drill can improve the ratio of crop yield (seeds harvested per seed planted) by as much as nine times. The use of seed drill saves time and labor.

Some machines for metering out seeds for planting are called planters. The concepts evolved from ancient Chinese practice and later evolved into mechanisms that pick up seeds from a bin and deposit them down a tube.

Seed drills of earlier centuries included single-tube seed drills in Sumer and multi-tube seed drills in China, and later a seed drill by Jethro Tull that was influential in the growth of farming technology in recent centuries. Even for a century after Tull, hand sowing of grain remained common.

Design

In older methods of planting, a field is initially prepared with a plow to a series of linear cuts known as *furrows*. The field is then seeded by throwing the seeds over the field, a method known as *manual broadcasting*. The seeds may not be sown to the right depth nor the proper distance from one another. Seeds that land in the furrows have better protection from the elements, and natural erosion or manual raking will cover them while leaving some exposed. The result is a field planted roughly in rows, but having a large number of plants outside the furrow lanes.

There are several downsides to this approach. The most obvious is that seeds that land outside the furrows will not have the growth shown by the plants sown in the furrow since they are too shallow on the soil. Because of this, they are lost to the elements. Many of the seeds remain on the surface where they are vulnerable to being eaten by birds or carried away on the wind. Surface seeds commonly never germinate at all or germinate prematurely, only to be killed by frost.

Since the furrows represent only a portion of the field's area, and broadcasting distributes seeds fairly evenly, this results in considerable wastage of seeds. Less obvious are the effects of overseeding; all crops grow best at a certain density, which varies depending on the soil and weather conditions. Additional seeding above this will actually reduce crop yields, in spite of more plants being sown, as there will be competition among the plants for the minerals, water, and the soil available. Another reason is that the mineral resources of the soil will also deplete at a much faster rate, thereby directly affecting the growth of the plants.

The invention of the seed drill dramatically improved germination. The seed drill employed a series of runners spaced at the same distance as the plowed furrows. These runners, or drills, opened the furrow to a uniform depth before the seed was dropped. Behind the drills were a series of presses, metal discs which cut down the sides of the trench into which the seeds had been planted, covering them over.

This innovation permitted farmers to have precise control over the depth at which seeds were planted. This greater measure of control meant that fewer seeds germinated early or late and that seeds were able to take optimum advantage of available soil moisture in a prepared seedbed. The result was that farmers were able to use less seed and at the same time experience larger yields than under the broadcast methods.

Uses

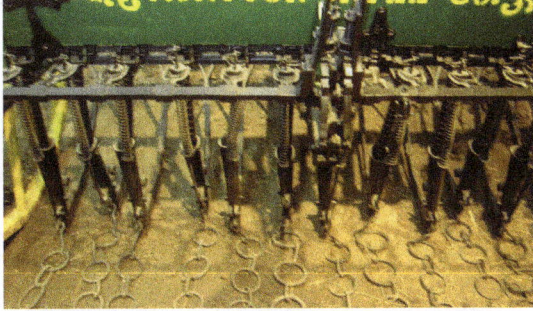

1902 model 12-run seed drill produced by Monitor Manufacturing\Company, Minneapolis, Minnesota.

Drilling is the term used for the mechanized sowing of an agricultural crop. Traditionally, a seed drill used to consist of a hopper filled with seeds arranged above a series of tubes that can be set at selected distances from each other to allow optimum growth of the resulting plants. Seeds are spaced out using fluted paddles which rotate using a geared drive from one of the drill's land wheels—seed rate is altered by changing gear ratios. Most modern drills use air to convey seed in plastic tubes from the seed hopper to the coulters—it is an arrangement which allows seed drills to be much wider than the seed hopper—as much as 12 m wide in some cases. The seed is metered mechanically into an air stream created by a hydraulically powered onboard fan and conveyed initially to a distribution head which sub-divides the seed into the pipes taking the seed to the individual colters.

Modern air seeder and hoe drill combination

The seed drill allows farmers to sow seeds in well-spaced rows at specific depths at a specific seed rate; each tube creates a hole of a specific depth, drops in one or more seeds, and covers it over. This invention gives farmers much greater control over the depth that the seed is planted and the ability to cover the seeds without back-tracking. The result is an increased rate of germination, and a much-improved crop yield (up to eight times).

The use of a seed drill also facilitates weed control. Broadcast seeding results in a random array of growing crops, making it difficult to control weeds using any method other than hand weeding. A field planted using a seed drill is much more uniform, typically in rows, allowing weeding with the hoe during the growing season. Weeding by hand is laborious and inefficient. Poor weeding reduces crop yield, so this benefit is extremely significant.

Before the operation of the seed drill, the ground must be plowed and harrowed. The plow digs up the earth and the harrow smooths the soil and breaks up any clumps. The drill must then be set for the size of the seed used. Afterwards, the grain is put in the hopper on top which then follows along behind the drill while it spaces and plants the seed. This system is still used today but has been modified and updated such that a farmer can plant many rows of seed at the same time.

A seed drill can be pulled across the field using bullocks or a tractor. Seeds sown using a seed drill are distributed evenly and placed at the correct depth in the soil.

Zero-till-ferti-seed-drill

A Zero-till-ferti-seed-drill machine has been developed at G.B.Pant University of Agriculture and Technology, Pantnagar by which direct sowing of wheat is done in Rice field without ploughing. This helps advancing the sowing of wheat as the time required for field preparation is saved. Zero-tillage can be adopted with following preparations:

- At the time of sowing there should be proper moisture in the field.

- Rice should be harvested near the ground and the left over stubble should not be more than 15 cm in height and field should be free from weeds.

- At the time of sowing the seed-drill should be lifted up or lower down very slowly to avoid clocking of furrow opener by soil, otherwise seeds and fertilizer will not drill in the furrow.

- Seed should be treated with vitavax or Bavistin at the rate of 2.5 g/kg of wheat seed. Seed rate should be 140-150 kg/ha (20-25% higher).

- Sowing depth should be maintained about 5-6 cm.

- Light planker may be used behind the zero-tillage machine.

- After sowing by Zero-till-seed-drill, other package of practices remain the same as in other methods.

The Zero-till-ferti-seed-drill has knife type lines in place of shovel type and is suitable for sowing under Zero-tillage conditions as well as conventional field preparation. The lines are fitted at a distance of 20 centimeters and have a provision to shift by

2.5 centimeters on either side. The machine is fitted with 2 boxes, one for seed and the other one for the fertilizer. On each box, lever is provided along with locking bolts and marking to get known quantity per hectare. Seedling depth can be manipulated by adjusting two side depth wheels with the help of screw bolts. Front driving wheel is provided with a groove to adjust as per requirement. A wooden platform is provided to keep a man to monitor the chocking of seed and fertilizer tube. Zero-tillage machines are provided with two lower link pins and a lift patti to attach the machine with tractor.

References

- Joseph Needham; Gwei-Djen Lu; Ling Wang (1987). Science and civilisation in China. Cambridge University Press. pp. 48–50. ISBN 978-0-521-30358-3.

- Dibbling, The-farmer's-and-planter's-encyclopaedia-of-rural-affairs: booksupstairs.com, Retrieved 30 April, 2019

- Hounshell, David A. (1984), From the American System to Mass Production, 1800-1932: The Development of Manufacturing Technology in the United States, Baltimore, Maryland: Johns Hopkins University Press, ISBN 978-0-8018-2975-8, LCCN 83016269

- Dibbling-method-with-advantage: cststudy.blogspot.com, Retrieved 10 July, 2019

- "Jeffers, D. L., & Beuerlein, J. (2001). Aerial and Other Broadcast Methods of Seeding Wheat" (PDF). osu.edu. Archived from the original (PDF) on 2003-08-03. Retrieved 2012-03-08

Diverse Aspects of Seed Science

The branch of science which is involved in the processing, testing and conservation of food and seeds is known as seed science. Some of the areas which are under focus in this discipline are seed testing, seed pelleting and seed storage. The chapter closely examines these key concepts and techniques of seed science to provide an extensive understanding of the subject.

SEED TESTING

Seed testing is science of evaluating the planting value of the seeds. By seed testing we can asses the quality attributes of the seed lots which have to be offered for sale and minimizing the risk of planting low quality seeds.

Objectives of Seed Testing

- To determine the seed quality i.e., their sustainability for planting.

- To identify seed quality problems and their probable causes.

- To determine if seed meets established quality standards or labeling specifications.

- To establish quality and provide a basis for price and consumer discrimination among lots in the market.

- To determine the need for drying and processing and specific procedures that should be used.

Important Components of Seed Testing

- Seed sampling

- Physical purity

- Germination

- Seed moisture

- Seed viability

- Seed health

- Seed Vigour Testing

Seed Sampling

Seed sampling is the process of obtaining a sample of required size for test in which the same constituents are present as in the seed lot and in same proportion. Types of samples received at STL:

- Service sample

- Official sample

- Charged sample

- Seed sampling

- Primary sample: a fraction of seeds taken randomly from seed lot different places.

- Composit sample: pooled sample of all primary samples.

- Submitted sample: sample which is submitted by producer to seed testing laboratory. It is large in size.

- Working sample: subdivision of the submitted sample.

- Equipments for sampling: Seed Triers-Stick or sleeve type trier, Bin sampler, Nobbe trier.

- Seed dividers: Boerner seed divider, Gamet seed divider, Soil type seed devider.

- Random cup method: Suitable for crops requiring sample up to 10g. Place 6-8 cups in the tray and pour seeds uniformly over tray and select the cup randomly.

- Spoon method: Permissible for sampling of small seeded seeds. Pour the seeds evenly over the tray. Remove small portion of seeds using spoon and spatula.

- Hand halving method: Restricted to specific genera of chaffy seeds. Pour the seeds on floor and make a mound. Mound is then divided into half and each half is halved again. The primary objective of purity analysis is to determine whether the submitted seed sample confirms to the prescribed quality standards in regard to purity components, objectionable weed seeds, other inseparable crop seeds and seeds of other distinguishable varieties.

Equipments and Materials

Purity table or working board, Seed divider, Forceps, Spatula, Brush Aluminum purity dish, Magnifier (5 to 7x), Analytical balance, Hand screen (with ISI specifications), Top loading balance (as mentioned above), Stereoscopic binocular microscope, Seed blower, table lamp, Watch glass.

Components of Seed

- Pure seed: Seeds of kind/species stated by sander or found to predominant in the test.
- Other crop seeds: seeds of plants which are as crops , other than main crop.
- Inert matter: includes seed units and all other matter and structures not defined as pure seed, other crop seed or weed seed.
- Weed seeds: Seeds of a weed species which are recognized as weeds by law/ general usage.

Reporting of Results

- Weight by percentage single decimal place.
- All components should add to 100 %.
- Less than 0.05% Reported as trace.
- Any component is found to be nil- reported as 0.0.
- The components scientific names should be mentioned.

Germination

It is the emergence and development from the seed embryo of those essential structures which for the kind of seeds being tested, indicate the ability to develop into a normal plant under favorable conditions in the soil objectives of the germination test:

- To know the field planting value of the seed
- Results can be used to compare among the seed lots

Materials Required

Sand (0.05 to 0.8 mm particle size)/Germination paper,Germinaton chamber, Plastic boxes for seed germination in sand, Rubber band Glass Marking Pencil, Miscellaneous laboratory glass wares different substrata paper media:

- Germination paper

- Blotter paper

- Pleated paper

Sand Media

Characteristics of Germination Paper

- It should be porous in nature

- It should have maximum water holding capacity to ensure continuous supply of water during the test period

- Free from bacteria, dirt, fungi and toxic substances

- Made out of 100%cellulose with

- pH should be 6-7.5

- Characteristics of germination paper

- Paper should posses sufficient strength to the prevent penetration of root in to the paper

- Paper size is 46 X 29cm

- It should have reasonable cost

- Should not serve as suitable media for saprophytic Fungi

Methods of Germination Test

- Between paper method

- Top of the paper method

- Sand method

According to ISTA 400 seeds, 100 seeds in each replication (4 repns) has to be used for seed germination test.

Spacing between seeds should be uniform and there should not be any over lapping of seeds. Temperature to that specific species should be maintained. Species which required light (Grasses) has to be provided.

Evaluation of Germination Test

- First count

- Final count

Categories of Seedlings

Normal Seedlings

It shows the potential for continued development in to satisfactory plant when grown in good quality soil and under favourable conditions of moisture, temperature and light.

Abnormal Seedlings

The potential for continued development in to satisfactory plant when grown in good quality soil and under favourable conditions of moisture, temperature and light.

Fresh Ungerminated Seeds

Seeds which have not germinated by end of test period but remain clean and firm.

Hard Seeds

Remain hard by end of test period.

Dead Seeds

Seeds which are neither hard nor fresh nor produced any part of seedlings at end of test period.

Seed Moisture Test

The main purpose of moisture determination of seeds is to prevent loss of seed viability till it is planted for commercial purposes.

Seed Moisture Estimation

Rapid method using moisture meters Hot air oven method rapid method using moisture meters:

- Electronic moisture meter
- Universal moisture meter
- Infrared moisture balance

Hot Air Oven Method

- Low constant temperature method (103 degree C for 17 hours).
- High constant temperature method (130 degree C for 2-3 hours).

Viability Test: Tz Test

Tz is a biochemical test and one of the quick methods to predict seed viability developed by Lakon in Germany.

Principle

In this biochemical test, living cells are made visible by reduction of an indicatordye. Indicator dye 2, 3, 5 Tripheny tetrazolium chloride (soluble, colourless and diffusible) interfeares with the reduction process of living cells and accepts hydrogen from the hydrogenases. Red, stable and nondiffusabe Formazon is produced in living cells. This makes it possible to distinguish the red coloured living parta of seeds from the colour less dead ones.

Procedure: Preparation of Seeds

- Bisect longitudinally: eg. Sorghum, large seeded grasses.
- Bisect laterlly: eg. Small seeded grasses.
- Remove seed cot: eg. Dicots.
- Condtioning only: eg: large seeded legumes.

Advantages of the Tz Test

- Rapid test
- More reliability
- Few equipments needed

Disadvantages of the Tz test

- More time required in seed preparation
- Skilled labors required
- Vague results

Seed Health Testing

The objective is to determine the health status of the seed sample i.e., whether the seeds are infected with fungi or not. Seed borne inoculum may give rise to spread of disease in field and reduce the commercial value of the crop. Further the seeds imported from out side source may introduce new pests and diseases into new areas. Seed health testing methods: Inspecting dry seeds Seedling symptom test, Agar plate method, blotter method etc.

Seed Vigour Tests

Seed Vigour is a sum total of a all the attributes which resulted in quick, uniform and early emergence of seedlings even in unfavourable environmental conditions The principle objective of a seed vigour test is to differentiate a range of quality levels, for example high, medium and low vigour seeds. Direct Vigour tests Indirect vigour tests Brick gravel tests (2-3mm size) Dry weight of seedlings Paper piercing test Speed of germination Seedling length measurement RQ test GADA test TZ test EC of seed leachates test.

SEED BALL

Seed balls, also known as "earth balls" or *nendo dango*, consist of a variety of different seeds rolled within a ball of clay, preferably volcanic pyroclastic red clay. Various additives may be included, such as humus or compost. These are placed around the seeds, at the center of the ball, to provide microbial inoculants. Cotton-fibres or liquefied paper are sometimes mixed into the clay in order to strengthen it, or liquefied paper mash coated on the outside to further protect the clay ball during sowing by throwing, or in particularly harsh habitats.

The technique for creating seed balls was rediscovered by Japanese natural farming pioneer Masanobu Fukuoka. The technique was also used, for instance, in ancient Egypt to repair farms after the annual spring flooding of the Nile. In modern times, during the period of the Second World War, this Japanese government plant scientist working in a government lab, Fukuoka, who lived on the mountainous island of Shikoku, wanted to find a technique that would increase food production without taking away from the land already allocated for traditional rice production which thrived in the volcanic rich soils of Japan.

Construction

To make a seed ball, generally about five measures of red clay by volume are combined with one measure of seeds. The balls are formed between 10mm and 80mm (about 0.4 to 3.15 inches) in diameter.

Seed Bombing

Seed bombing is the practice of introducing vegetation to land by throwing or dropping seed balls. It was made popular by green movements such as guerrilla gardening as a way to practice active reforestation.

Seed balls were also experimentally used in aerial seeding in Kenya in 2016. This was an attempt to improve the yield of standard aerial seeding.

Aerial seeding (or aerial reforestation) is the technique of spreading seeds from an areoplane, an helicopter or a similar flying transport. It can be considered a particular type of direct seeding: as such, it introduces seeds directly in the field and it is often not economical due to the issues of germination, pests and seed predation by rodents or other wild animals. Transplanting seedlings from a plant nursery to the field is a more effective sowing technique.

Aerial seeding has a low yield and require 25% to 50% more seeds than drilled seeding to achieve the same results. It is sometimes used as a technique to plant cover crops without having to wait for the main crop's off-season. The earliest attempts at aerial reforestation date back to the 1930s. In this period, planes were used to distribute seeds over certain inaccessible mountains in Honolulu after forest fires. These experiments were largely unsuccessful, because of poor seed dispersal: seeds failed to obtain enough kinetic energy to enter the soil and as a result were massively predated. This in turn generated an infestation of rodents in Hawaii.

The use of seed balls, instead of simple seeds, to perform aerial reforestation in Kenya seems to have produced desirable results. Chardust Ltd, the company involved and distributing seed balls for that project, claims to have sold and distributed over 7 million seedballs, as per August 2019. It is likely, though, that the majority of these seed balls are deployed traditionally instead than via aerial seeding and there is no published data to support the benefits of using seed balls via aerial seeding.

In 1987, Lynn Garrison proposed the creation of a Haitian Aerial Reforestation Project (HARP), by which tons of seed would be scattered from specially modified aircraft. The seeds would be encapsulated in an absorbent material. This coating would contain fertilizer, insecticide/animal repellent and, perhaps a few vegetable seeds. Haiti has a bimodal rainy season, with precipitation in spring and fall. The seeds could have been moistened a few days before the drop, to start germination. Unfortunately the project never came to fruition.

Another more recent project idea was to use saplings instead of seeds in aerial. Saplings would be encased in sturdy and biodegradable containers which would work as projectiles and pierce the ground at high velocity. This would likely guarantee a better yield when compared to simple aerial seeding or even seed bombing.

This project was being developed in 1999 by a company called Aerial Reforestation Inc, in Newton, Massachusetts, based on an original idea by pilot Jack Walters. The company was planning to use military transport aircraft C-130, traditionally used to lay out landmines on combat fields. As per 2019 the company does not seem to be operating anymore. Other researchers are still investigating the potential of these "aerial sapling darts", by improving their aerodynamics to achieve better soil penetration and therefore higher reforestation yields. More research is needed to assess exactly their performance against other reforestation methods.

Feasibility and Weaknesses

Despite its low yield, interest in aerial seeding has recently grown, while searching for a quick system to obtain reforestation and fight global warming. The advantage of using an airplane/helicopter is the ability to quickly seed large areas, even remote areas, otherwise impractical to be used in active reforestation.

Aerial seeding is therefore best suited to sites whose remoteness, ruggedness, inaccessibility, or sparse population make seedling planting difficult. It is particularly appropriate for "protection forests" because helicopters or planes can easily spread seed over steep slopes or remote watersheds and isolated dryland areas. It seems also well suited for use in areas where there may be a dearth of skilled laborers, supervisors, and funds for reforestation. It has the potential to help increase production of tree crops for forage, food, and honey as well as wood for fuel, posts, lumber, charcoal and pulp.

Seed balls and aerial reforestation are heavily sensitive to environmental conditions. Seeds deployment may not be practical in some cases because the site may require preparation or the season may be wrong. To germinate successfully, seeds usually must fall directly onto mineral soil rather than onto established vegetation or undecomposed organic matter. Where organic matter has accumulated thickly, the site must normally be burned, furrowed, or disked. The soil disturbance left after logging is often sufficient. Rough terrain is more amenable to broadcast seeding, but yields are typically low even in the best conditions.

On certain sites ground preparation may be necessary. Site preparation and the seeding operation must be well coordinated to meet the biological requirements for prompt seed germination and seeding survival. Dry sites may have to be specially ridged or disked so as to optimize the rainfall that reaches the seed. Excessively wet sites may need to be ridged or drained.

The degree of field slope is not critical as long as seeds find a receptive seedbed. Steep watersheds, eroding mountain slopes, bare hillsides, and spoil-banks where vegetation is sparse are often suitable for aerial seeding (however, on some steep slopes with smooth, bare soil, rain may wash the seeds away too easily for successful seeding).

Arid and savanna lands (for example, those where annual rainfall is under 800 mm) are most in need of reforestation. These are regions where aerial seeding in principle has exceptional potential. They include vast tracts of unused or poorly used land that has sparse tree cover and that is not confined to private land holdings, so it is generally accessible to aircraft. The native trees (such as species of Acacia, and other genera) in these areas are generally well adapted for survival under difficult field conditions. These are not species for timber as much as for firewood, forage, fruit, gum, erosion control, and other such uses.

As a prerequisite to any method of reforestation, the species selected must be adapted to the temperature, length of growing season, rainfall, humidity, photoperiod, and other environmental features of the area. Ideally, before aerial seeding takes place trial plots should be established to test those species most likely to germinate and grow successfully on the chosen sites. Even when one species has the right characteristics, it may be prudent to test seed of different provenances to find those best suited to the site.

Characteristics that make a particular species more or less appropriate for aerial seeding include:

- Seed size;

- Seed availability;

- Ability of the seed to germinate on the soil surface;

- Germination and seedling growth speeds;

- Ability to withstand temperature extremes and prolonged dry periods (orthodox seed);

- Ability to tolerate soil conditions;

- Light tolerance;

- Seed stability when stored in large quantities;

- Suitability of seed for handling with mechanical seeding devices;

- Development speed of a deep taproot by seedlings to enable them to withstand adverse climatic conditions in the period following germination.

Species with highly palatable seeds have little prospect of success because wildlife eat the seed before it has a chance to germinate unless it is pelletized. Also, small seeds and lightweight, chaffy seeds are more likely to drift in the wind, so they are harder to target during the drop. Small seeds, however, fall into crevices and are then more likely to get covered with soil, thereby enhancing their chances of survival. Aerial seeding may prove to work best with "pioneer" species, which germinate rapidly on open sites, are adapted for growth on bare or disturbed areas, and grow well in direct sunlight.

Aerial Seeding Deployment Methods

- Crop Spraying Aircraft. It is the most common method and the one practiced by Farmland Aviation, Kenya, one of the few companies active in this field. They claim to be able to spread up to six tons of tree seeds per hour over tens of thousands of acres.

- Unmanned Aerial Vehicles: Drones are not presently used in aerial seeding. Current day low-cost UAV's lack of payload capacity and range limits them for most aerial seeding applications. They could be employed in the future if specifications improve.

- Paragliding: This deployment method is being tested in Kenya and holds promises of great reforestation rates because of low cost, low speed and altitudes, even if seed spraying rate is likely to be much slower than deployment by aircraft.

SEED STORAGE

Seed Storage is the preservation of seed with initial quality until it is needed for planting. The ability of seed to tolerate moisture loss allows the seed to maintain the viability in dry state. Storage starts in the mother plant itself when it attains physiological maturity. After harvesting the seeds are either stored in ware houses or in transit or in retail shops. During the old age days, the farmers were used farm saved seeds, in little quantity, but introduction of high yielding varieties and hybrids and modernization of agriculture necessitated the development of storage techniques to preserve the seeds.

Objective of Seed Storage

To maintain initial seed quality viz., germination, physical purity, vigour etc., all along the storage period by providing suitable or even better conditions. Since the main objective of seed storage is maintenance of an acceptable capacity for germination and emergence, it can only be accomplished by reducing the rate ofdeterioration to the degree required to maintain an acceptable level of quality for the desired period.

Purpose of Seed Storage

Seed storage is the maintenance of high seed germination and vigour form harvest until planting. Is important to get adequate plant stands in addition to healthy and vigourous plants. Every seed operation has or should have a purpose. The purpose of seed storage is to maintain the seed in good physical and physiological condition from the time they are harvested until the time they are planted. Seeds have to be stored, of course, because there is usually a period of time between harvest and planting. During this period, the seed have to be kept somewhere. While the time interval between harvest and planting is the basic reason for storing seed, there are other considerations, especially in the case of extended storage of seed.

Seed suppliers are not always able to market all the seed they produce during the following planting season. In many cases, the unsold seed are "carried over" in storage for marketing during the second planting season after harvest. Problems arise in connection with carryover storage of seed because some kinds, varieties, and lots of seed do not carryover very well.

Seeds are also deliberately stored for extended periods so as to eliminate the need to produce the seed every season. Foundation seed units and others have found this to be an economical, efficient procedure for seed of varieties for which there is limited demand. Some kinds of seed are stored for extended periods to improve the percentage and rapidity of germination by providing enough time for a "natural" release from dormancy.

Regardless of the specific reasons for storage of seed, the purpose remains the same maintenance of a satisfactory capacity for germination and emergence. The facilities and procedures used in storage, therefore, have to be directed towards the accomplishment of this purpose.

In the broadest sense the storage period for seed begins with attainment of physiological maturity and ends with resumption of active growth of the embryonic axis, i.e., germination. Seeds are considered to be physiologically and morphologically mature when they reach maximum dry weight. At this stage dry-down or dehydration of the seed is well underway. Dry-down continues after physiological maturity until moisture content of the seed and fruit decreases to a level which permits effective and efficient harvest and threshing. This stage can be termed as harvest maturity. There usually is an interval of time between physiological maturity and harvestable maturity, and this interval represents the first segment of the storage period. Any delays in harvesting the seed after they reach harvest maturity prolongs the first segment of the storage period – often to the detriment of seed quality.

The second segment of the storage period extends from harvest to the beginning of conditioning. Seed in the combine, grain wagon, and bulk storage or drying bins are in storage and their quality is affected by the same factors that affect the quality of seed during the packaged seed segment of the storage period. The third segment of the storage period begins with the onset of conditioning and ends with packaging. The fourth segment of the storage period is the packaged seed phase The packaged seed segment is followed by storage during distribution and marketing, and finally by storage on the farm before and during planting.

The control that a seedsman has over the various segments of the storage period for seed varies from a high degree of control from harvest to distribution, to much less control during the postmaturation-preharvest, distribution-marketing, and on-farm segments. Despite variable degrees of control over the various segments of the storage period, the seedsman's plans for storage must take into consideration all the segments. The things that can be done must be done if the quality of the seed is to be maintained.

Seed Bank

A seed bank (also seedbank or seeds bank) stores seeds to preserve genetic diversity; hence it is a type of gene bank. There are many reasons to store seeds. The genes that plant breeders need to increase yield, disease resistance, drought tolerance, nutritional quality, taste, etc. of crops. Another is to forestall loss of genetic diversity in rare or imperiled plant species in an effort to conserve biodiversity ex situ. Many plants that were used centuries ago by humans are used less frequently now; seed banks offer a way to preserve that historical and cultural value. Collections of seeds stored at constant low temperature and low moisture are guarded against loss of genetic resources that are otherwise maintained in situ or in field collections. These alternative "living" collections can be damaged by natural disasters, outbreaks of disease, or war. Seed banks are considered seed libraries, containing valuable information about evolved strategies to combat plant stress, and can be used to create genetically modified versions of existing seeds. The work of seed banks spans decades and even centuries. Most seed banks are publicly funded and seeds are usually available for research that benefits the public.

Storage Conditions and Regeneration

Seeds are living creatures and keeping them viable over the long term requires adjusting storage moisture and temperature appropriately. As they mature on the mother plant, many seeds attain an innate ability to survive drying. Survival of these so-called 'orthodox' seeds can be extended by dry, low temperature storage. The level of dryness and coldness depends mostly on the longevity that is required and the investment in infrastructure that is affordable. Practical guidelines from a US scientist in the 1950s and 1960s, James Harrington, are known as 'Thumb Rules.' The 'Hundreds Rule' guides that the sum of relative humidity and temperature (in Fahrenheit) should be less than 100 for the sample to survive 5 years. Another rule is that reduction of water content by 1% or temperature by 10 degrees Fahrenheit will double the seed life span. Research from the 1990s showed that there is a limit to the beneficial effect of drying or cooling, so it must not be overdone.

Understanding the effect of water content and temperature on seed longevity, the Food and Agriculture division of the United Nations and a consultancy group called Bioversity International developed a set of standards for international seed banks to preserve seed longevity. The document advocates drying seeds to about 20% relative humidity, sealing seeds in high quality moisture-proof containers, and storing seeds at -20 degrees Celsius. These conditions are frequently referred to as 'conventional' storage protocols. Seeds from our most important species - corn, wheat, rice, soybean, pea, tomato, broccoli, melon, sunflower, etc. can be stored in this way. However, there are many species that produce seeds that do not survive the drying or low temperature of conventional storage protocols. These species must be stored cryogenically. Seeds of citrus fruits, coffee, avocado, cocoa, coconut, papaya, oak, walnut and willow are a few examples of species that should be preserved cryogenically.

Like everything, seeds eventually degrade with time. It is hard to predict when seeds lose viability and so most reputable seed banks monitor germination potential during storage. When seed germination percentage decreases below a prescribed amount, the seeds need to be replanted and fresh seeds collected for another round of long-term storage.

Challenges

- Knowing what to store in a seed bank is the greatest challenge. Collections must be relevant and that means they must provide useful genetic diversity that is accessible to the public. Collections must also be efficient and that means they mustn't duplicate materials already in collections.

- Keeping seeds alive for hundreds of years is the next biggest challenge. Orthodox seeds are amenable to 'conventional' storage protocols but there are many seed types that must be stored using nonconventional methods. Technology for these methods is rapidly advancing; local institutional infrastructure may be lacking.

Alternatives

In-situ conservation of seed-producing plant species is another conservation strategy. In-situ conservation involves the creation of National Parks, National Forests, and National Wildlife Refuges as a way of preserving the natural habitat of the targeted seed-producing organisms. In-situ conservation of agricultural resources is performed on-farm. This also allows the plants to continue to evolve with their environment through natural selection.

- An arboretum stores trees by planting them at a protected site.

- A less expensive, community-supported seed library can save local genetic material.

The phenomenon of seeds remaining dormant within the soil is well known and documented. Detailed information on the role of such "seed banks" in northern Ontario, however, is extremely limited, and research is required to determine the species and abundance of seeds in the soil across a range of forest types, as well as to determine the function of the seed bank in post-disturbance vegetation dynamics.

Longevity

Seeds may be viable for hundreds and even thousands of years. The oldest carbon-14-dated seed that has grown into a viable plant was a Judean date palm seed about 2,000 years old, recovered from excavations at the palace of Herod the Great in Israel.

In February 2012, Russian scientists announced they had regenerated a narrow leaf campion (Silene stenophylla) from a 32,000-year-old seed. The seed was found in a burrow 124 feet (38 m) under Siberian permafrost along with 800,000 other seeds.

Seed tissue was grown in test tubes until it could be transplanted to soil. This exemplifies the long-term viability of DNA under proper conditions.

Climate Change

Conservation efforts such as seed banks are expected to play a greater role as climate change progresses. Seed banks offer communities a source of climate-resilient seeds to withstand changing local climates. As challenges arise from climate change, community based seed banks can improve access to a diverse selection of locally adapted crops while also enhancing indigenous understandings of plant management such as seed selection, treatment, storage, and distribution.

Facilities

Plant tissue cultures being grown at a USDA seed bank, the National Center for Genetic Resources Preservation.

There are about 6 million accessions, or samples of a particular population, stored as seeds in about 1,300 genebanks throughout the world as of 2006. This amount represents a small fraction of the world's biodiversity, and many regions of the world have not been fully explored.

- The Svalbard Global Seed Vault has been built inside a sandstone mountain in a man-made tunnel on the frozen Norwegian island of Spitsbergen, which is part of the Svalbard archipelago, about 1,307 kilometres (812 mi) from the North Pole. It is designed to survive catastrophes such as nuclear war and world war. It is operated by the Global Crop Diversity Trust. The area's permafrost will keep the vault below the freezing point of water, and the seeds are protected by 1-metre thick walls of steel-reinforced concrete. There are two airlocks and two blast-proof doors.

- The Millennium Seed Bank housed at the Wellcome Trust Millennium Building (WTMB), located in the grounds of Wakehurst Place in West Sussex, near London, in England, UK. It is the largest seed bank in the world (longterm, at least 100 times bigger than Svalbard Global Seed Vault), providing space for the storage of billions of seed samples in a nuclear bomb proof multi-story underground vault. Its ultimate aim being to store every plant species possible, it reached its first milestone of 10% in 2009, with the next 25% milestone aimed to be reached by 2020. Importantly they also distribute seeds to other key locations around the world, do germination tests on each species every 10 years, and other important research.

- The Australian Grains Genebank (AGG) is a national center for storing genetic material for Plant breeding and research. The Genebank is in a collaboration with the Australian Seed Bank Partnership on an Australian Crop Wild Relatives project. It is located at Grains Innovation Park, in Horsham, Victoria, Australia, and was officially opened in March, 2014 The primary reason for the bank to be created was the extreme temperatures in the area, up to 40 degrees Celsius (104 degrees Fahrenheit) in the summer time. Because of that they had to ensure the protection of the grains all year around. The Genebank aims to collect and conserve the seeds of Australian crop wild species, that are not yet adequately represented in existing collections.

- The former NSW Seedbank focuses on native Australian flora, especially NSW threatened species. The project was established in 1986 as an integral part of The Australian Botanic Gardens, Mount Annan. The NSW Seedbank has collaborated with the Millennium Seed Bank since 2003. The seed bank has since been replaced as part of a major upgrade by the Australian Plant-Bank.

- Nikolai Vavilov was a Russian geneticist and botanist who, through botanic-agronomic expeditions, collected seeds from all over the world. He set up one of the first seed banks, in Leningrad (now St Petersburg), which survived the 28-month Siege of Leningrad in World War II. It is now known as the Vavilov Institute of Plant Industry. Several botanists starved to death rather than eat the collected seeds.

- The BBA (Beej Bachao Andolan — Save the Seeds movement) began in the late 1980s in Uttarakhand, India, led by Vijay Jardhari. Seed banks were created to store native varieties of seeds.

- National Center for Genetic Resources Preservation, Fort Collins, Colorado, United States.

- Desert Legume Program (DELEP) focuses on wild species of plants in the legume family (Fabaceae), specifically legumes from dry regions around the world. The DELEP seed bank currently has over 3600 seed collections representing nearly 1400 species of arid land legumes originating in 65 countries on six continents. It is backed up (at least in part) in National Center for Genetic Resources Preservation, and in the Svalbard Global Seed Vault. The DELEP seed bank is an accredited collection of the North American Plant Conservation Consortium.

Soil Seed Bank

Soil seed bank is a natural storage of seeds in the leaf litter, on the soil surface, or in the soil of many ecosystems, which serves as a repository for the production of subsequent generations of plants to enable their survival. The term soil seed bank can be used to describe the storage of seeds from a single species or from all the species in a particular area. Given the variety of stresses that ecosystems experience—such as cold, wildfire, drought, and disturbance—seed banks are often a crucial survival mechanism for many plants and maintain the long-term stability of ecosystems.

The Role of Seed Dormancy

Seed dormancy and environmental constraints on germination influence various characteristics of soil seed banks. For example, seed dormancy determines how long a seed can remain viable in the soil. Factors such as embryo immaturity, chemical inhibitors, and physical constraints influence seed dormancy. Light filtered through plant canopies, for example, can inhibit germination in some species, while a long winter chilling may break dormancy in other species. The result is a considerable variety in the patterns of germination of the seed banks by seasons, disturbances, or other environmental shifts.

Variation in the characteristics of seed dormancy determine whether a species's soil seed bank is transient (temporary) or persistent. Transient seed banks are composed of species that produce seeds with a brief or no period of dormancy. Such seeds generally germinate prior to the next round of seed production, and the seed bank is thus continually depleted and reestablished. Transient seed banks are typical for many plants, especially long-lived perennials such as trees and shrubs. Often, such species rely on other strategies or life-history stages for persistence. For example, species may depend on long-lived adults, "banks" of seedlings in a forest understory, or extensive seed dispersal. In contrast, species with persistent seed banks have seeds that can remain dormant for more than a year, meaning that there is always some viable seed in the soil as a reserve. Persistent seed banks are common in annual plants and some woody plants, in which the failure of seed to establish the next generation would mean the collapse of the population. Scientists sometimes further classify persistent seed banks based on the extent or pattern of dormancy.

The Role of Disturbance

In addition to dormancy, considerable variation occurs in seed bank germination because of seasonal or other environmental shifts. Disturbances such as fire, flooding, windstorms, plowing, or forest clearing are frequently strong selective forces and may increase the overall germination response of seeds. Ecosystems characterized by wildfire often have extreme cases of persistent seed banks, as is common for many areas with Mediterranean climates, such as Australia, California, and South Africa. In those ecosystems the germination of many species requires signals provided by fire, such as a heat pulse into the soil or chemicals from smoke or charred wood. Germination may not occur until after a wildfire, which then results in mass germination from the seed bank the following spring. Similarly, the seed banks of agricultural weeds are often well adapted to the almost continuous human-made disturbances of their environment. Such weeds frequently have complex dormancy patterns that reflect the agricultural practices under which they evolved.

Seed Bank Modeling

Researcher Dan Cohen was one of the first scientists to model soil seed banks. In the 1960s, focusing on desert annuals subject to highly irregular rainfall, he developed population-dynamics models that suggested that a reserve of some fraction of seed in the soil was essential for the plants to avoid local extinction. Cohen found that the dynamics of soil seed banks reflect the degree of ecological constraint a species or population faces in establishing the next generation. Although his work focused on annuals, the conceptual framework applies readily to any plant species. Such modeling is important to ecological research and conservation planning, as traditional demographic models and field surveys often fail to consider population reserves in the soil.

Importance of Seed Banks

Preservation of Crop Diversity

This is the most important reason for the storage of seeds. Just as human beings and animals are adapted to different conditions for survival, so are crops. Different types of the same species exist due to this adaptive nature. Therefore, it is of critical necessity that such diversity is preserved.

Protection from Climate Change

For a couple of decades now, the world has witnessed radical climatic change that has been accelerated by increased industrial pollution. Crop extinction is inevitable with such extreme changes. If seeds are stored in seed banks, the danger of total elimination of certain species of crops is eliminated.

Protection from Natural Disasters

Natural disasters are unforeseen events that could lead to complete annihilation of crops from the face of the earth. The foresight of keeping seeds in a seed bank could save such a situation. Malaysian rice paddies, for example, were wiped out during the 2004 tsunami and international seed banks provided farmers with seeds that helped them start over.

Disease Resistance

Crop diseases are highly contagious and very deadly to plants. A serious breakout could completely eliminate crops. Where diseases have ravaged crops and left no traces that farmers could start on, seed banks can intervene and provide them with seeds that will enable them start on a clean slate.

Provide Seed Material for Research

Seeds that are stored in seed banks can be made easily available to scientists and researchers who wish to study these seeds especially if such research could lead to improvement of crop production.

Preservation from Man-made Disasters

Man-made disasters such as war and oil spills could lead to the annihilation of crops. Counties that are engaged in war make it difficult for farmers to continue farming and it's easy for crops to disappear. Once peace is restored, seeds can be retrieved from seed banks and replanted.

Properly stored seeds can stay viable for even millennia, eliminating the risk of losing crops that are critical for the existence of human beings and animals.

SEED PELLETING

Seed pelleting is the process of adding inert materials to seeds increasing their weight, size and shape. This improves plantability allowing for precise metering, spacing and depth of seed in the field.

If you're planting a crop that grows from small or irregularly shaped seeds, such as lettuce, carrots and onions, pelleting can help make planting significantly easier on you and your equipment. Pelleting turns a long thin seed into a larger, round-shaped seed, so seeds can be mechanically singulated much more readily and so placed accurately in the field. This helps place seeds precisely, which is a great advantage for crops like onion, which need consistent uniform planting distances between each seed to generate uniform bulb development.

We pellet seeds by applying solid particle fillers to the seed using a binder or adhesive. Seed coating pans are a derivation of confectionary pans. The seed tumbles gently in the pan and like a snowball becomes increasingly heavier, rounder and larger as pelleting material and adhesives are added. The spherical shape allows for precise seed singulation within the planters. Pelleting significantly changes the original size, shape, and weight of a seed. The weight can increase from 1000% to 4000% (a 10:1 to 40:1 ratio).

Types of Seeds which are Pelleted

Seeds of various sizes are commercially pelleted, from relatively large seeds like onion and tomato to very small seeds like lettuce species.

For onion, the seed can increase in weight 6-fold due to pelleting; there are approximately 230 raw seeds per gram, and after pelleting the diameter may be 13.5/64th of an inch (0.54cm). The volume for 1000 propagules is 3.7 cm³ for raw seed compared to 18.0 cm³ after pelleting.

Begonia is the smallest seed that Seed Dynamics pellets. Median seed weight for raw begonia is 88,000 seeds per gram. After pelleting, the seed count can average 857 seeds per gram, an increased mass of over 100-fold.

Ideal for Mechanical Metering of Valuable Seed

Pelleting is ideal for mechanical seed metering/planting in direct field and protected culture applications. In the coastal and desert valleys of California and southwest Arizona (as in other agricultural areas), it is common to use precision belt, plate and vacuum planters.

Split Pellet Technology for Increased O_2 Availability

Historically, the primary obstacle for successful seed pellet development has been slow and erratic germination associated with insufficient oxygen available to the seed. The development of a splitting pellet like our High-Density, Medium-Density or Light-Density lettuce pellets are especially beneficial to growers that plant lettuce under saturated soil-water conditions caused by frequent irrigation after sowing. A pellet that can split open upon hydration allows oxygenated water to move directly to the seed.

References

- Hong, T.D. and R.H. Ellis. 1996. A protocol to determine seed storage behaviour. IPGRI Technical Bulletin No. 1. (J.M.M. Engels and J. Toll, vol. eds.) International Plant Genetic Resources Institute, Rome, Italy. ISBN 92-9043-279-9

- Soil-seed-bank, science: britannica.com, Retrieved 18 May, 2019

- Drori, Jonathan (May 2009). "Why we're storing billions of seeds". TED2009. TED (conference). Retrieved 2011-12-11

- Seed-banking-benefits: permaculturenews.org, Retrieved 28 April, 2019

- Rajasekharan, P. E. (2015-01-01). "Gene Banking for Ex Situ Conservation of Plant Genetic Resources". In Bahadur, Bir; Rajam, Manchikatla Venkat; Sahijram, Leela; Krishnamurthy, K. V. (eds.). Plant Biology and Biotechnology. Springer India. pp. 445–459. doi:10.1007/978-81-322-2283-5_23. ISBN 9788132222828

- Pelleting, techniques, technology: seeddynamics.com, Retrieved 05 August, 2019

PERMISSIONS

We would like to thank the editorial team for lending their expertise to make the book truly unique. They have played a crucial role in the development of this book. Without their invaluable contributions this book wouldn't have been possible. They have made vital efforts to compile up to date information on the varied aspects of this subject to make this book a valuable addition to the collection of many professionals and students.

This book was conceptualized with the vision of imparting up-to-date and integrated information in this field. To ensure the same, a matchless editorial board was set up. Every individual on the board went through rigorous rounds of assessment to prove their worth. After which they invested a large part of their time researching and compiling the most relevant data for our readers.

The editorial board has been involved in producing this book since its inception. They have sent rigorous hours researching and exploring the diverse topics which have resulted in the successful publishing of this book. They have passed on their knowledge of decades through this book. To expedite this challenging task, the publisher supported the team at every step. A small team of assistant editors was also appointed to further simplify the editing procedure and attain best results for the readers.

Apart from the editorial board, the designing team has also invested a significant amount of their time in understanding the subject and creating the most relevant covers. They scrutinized every image to scout for the most suitable representation of the subject and create an appropriate cover for the book.

The publishing team has been an ardent support to the editorial, designing and production team. Their endless efforts to recruit the best for this project, has resulted in the accomplishment of this book. They are a veteran in the field of academics and their pool of knowledge is as vast as their experience in printing. Their expertise and guidance has proved useful at every step. Their uncompromising quality standards have made this book an exceptional effort. Their encouragement from time to time has been an inspiration for everyone.

The publisher and the editorial board hope that this book will prove to be a valuable piece of knowledge for students, practitioners and scholars across the globe.

INDEX